The Swedish Nuclear Dilemma

The Swedish Nuclear Dilemma

Energy and the Environment

William D. Nordhaus

Resources for the Future
Washington, DC

Printed in the United States of America

Published by Resources for the Future
1616 P Street, NW, Washington, DC 20036–1400

Library of Congress Cataloging-in-Publication Data

Nordhaus, William D.
The Swedish nuclear dilemma: energy and the environment / by William D. Nordhaus.
p. cm.
Includes bibliographical references and index.
ISBN 0–915707–84–5 (cloth)

1. Nuclear industry—Sweden. 2. Nuclear facilities—Sweden—Decommissioning. 3. Nuclear power plants—Environmental aspects—Sweden. 4. Energy policy—Sweden. 5. Environmental policy—Economic aspects—Sweden. I. Title.
HD9698.S82N67 1997
333.792'4'09485—dc21 97–19258
CIP

The paper in this book meets the guidelines for permanence and durability of the Committee on Production Guidelines for Book Longevity of the Council on Library Resources.

This book was typeset in Palatino by Betsy Kulamer and its cover was designed by AURAS Design.

RESOURCES FOR THE FUTURE (RFF) is an independent nonprofit organization engaged in research and public education on natural resources and environmental issues. Its mission is to create and disseminate knowledge that helps people make better decisions about the conservation and use of their natural resources and the environment. RFF takes responsibility for the selection of subjects for study and for the appointment of fellows, as well as for their freedom of inquiry. RFF neither lobbies nor takes positions on current policy issues.

Because the work of RFF focuses on how people make use of scarce resources, its primary research discipline is economics. Supplementary research disciplines include ecology, engineering, operations research, and geography, as well as many other social sciences. Staff members pursue a wide variety of interests, including the environmental effects of transportation, environmental protection and economic development, Superfund, forest economics, recycling, environmental equity, the costs and benefits of pollution control, energy, law and economics, and quantitative risk assessment.

Acting on the conviction that good research and policy analysis must be put into service to be truly useful, RFF communicates its findings to government and industry officials, public interest advocacy groups, nonprofit organizations, academic researchers, and the press. It produces a range of publications and sponsors conferences, seminars, workshops, and briefings. Staff members write articles for journals, magazines, and newspapers, provide expert testimony, and serve on public and private advisory committees. The views they express are in all cases their own and do not represent positions held by RFF, its officers, or trustees.

Established in 1952, RFF derives its operating budget in approximately equal amounts from three sources: investment income from a reserve fund; government grants; and contributions from corporations, foundations, and individuals. (Corporate support cannot be earmarked for specific research projects.) Some 45 percent of RFF's total funding is unrestricted, which provides crucial support for its foundational research and outreach and educational operations. RFF is a publicly funded organization under Section 501(c)(3) of the Internal Revenue Code, and all contributions to its work are tax deductible.

Contents

Foreword

One of the things that makes life both very frustrating and also very interesting is that accomplishing one objective frequently means backpedaling on another. Since economics is the study of tradeoffs, this means that there is generally plenty for economists to do. William Nordhaus is one of the best economists anywhere, and he has written a wonderful book about the tradeoffs faced by one country—Sweden—if and as it acts on a decision its citizens made in 1980 to phase out the use of nuclear power there. I should add that this decision has been reaffirmed by the Swedish Parliament on several occasions since the 1980 referendum, though with some elusive qualifications.

What will be both the environmental and also the economic implications of a Swedish phaseout of the use of nuclear power to generate electricity there? These are the two issues Nordhaus addresses in this book. The environmental tradeoffs are especially juicy. For at the same time that Sweden moves ahead to reduce its reliance on nuclear power, it is also committing itself to reducing emissions of carbon dioxide and other greenhouse gases. Yet the most natural substitutes for nuclear-powered electricity generation are boilers fueled by coal, fuel oil, or natural gas, any of which would add to, not reduce, Sweden's emissions of carbon dioxide. Throw into this mix a legislative commitment there not to expand that country's use of hydroelectric power—which currently accounts for more than half the electricity generated there—and you have on your hands a very thorny problem.

To answer the questions that he poses for himself, Nordhaus does not content himself with casual speculation. Rather, in the best academic tradition, he constructs a model of the Swedish electricity system ("SEEP," for Swedish Energy and Environmental Policy). He then uses this model to ask and answer such questions as: Where will replacement power come from if and when the existing nuclear reactors are shut down? How much will it cost? What effect will these costs have on the performance of the Swedish economy (which has not been setting the world on fire as it is)? How might Sweden reduce its greenhouse gas emissions while phasing out a very large source of electricity that is also carbon-free? And, finally—How will all this play out against the backdrop of electricity deregulation that is going on not only in Sweden but also across all of Europe and much of the rest of the world?

We at Resources for the Future are especially pleased to be publishing this interesting and important book. Not only is the analysis first rate and the writing clear and concise, but the subject could not fit better with RFF's ongoing research agenda. For the past two years, we have devoted as much or more attention RFF-wide to electricity deregulation and to global climate change as to any other subject. And we do so at a time when RFF's researchers are increasingly interested in the international dimensions of energy and environmental policy. Anyone interested in the nexus of these important topics could not but benefit from reading Bill Nordhaus's interesting book. I invite you to dig in.

Paul R. Portney
President
Resources for the Future

Preface

The present study was undertaken with the support and encouragement of SNS Förlag, a Swedish research organization housed in Stockholm. I am indebted to many people for assistance in the preparation of this report. I am particularly grateful to Stefan Lundgren for his infinite patience and profound understanding of the Swedish economy and to Lars Bergman for his tutoring in Swedish economic and energy matters. Åsa Ahlin was a critical part of the path in finding and synthesizing information. Nancy Jacobson, David Lam, and Kathy Merola helped out at the Yale end. I am grateful to Bo Andersson and Erik Hådén of the Stockholm School of Economics for their consultation and assistance in developing data on the Swedish energy system. Thanks go to a group of knowledgeable Swedish experts who served as a Reference Group at SNS, particularly its chairman, Carl-Johan Åberg, for comments and suggestions about the course of the study. I have profited from discussions with too many people to name in the Swedish government and energy industry. Finally, I am grateful to SNS for its unflagging encouragement in supporting this study. All errors and opinions contained in the report are the sole responsibility of the author.

I am particularly grateful for the encouragement and support on the western side of the Atlantic from Resources for the Future. Paul Portney conveyed both his valuable substantive suggestions and great enthusiasm, which made this publication possible. Rich Getrich not only gave his editorial blessing but also directed the project. Eric Wurzbacher managed the editorial and production phase and, along with Betsy Kulamer, Joann Platt, Kay Murphy, and Brian Kropp, turned a mountain of paper into a handsome volume.

The Swedish version of this study was published as William D. Nordhaus, *Kärnkraft och Miljö—Ett Svenskt Dilemma*, SNS Förlag, Stockholm, 1995.

William D. Nordhaus
Department of Economics
Yale University

PART I

Background

1

Introduction: The Swedish Dilemma

Half a century ago, the Swedish economist Gunnar Myrdal visited the United States and published his reflections on American society in *An American Dilemma*.[1] Today, as an American, I have been asked to bring an outsider's perspective to a central economic, political, and philosophical debate facing Sweden. This "Swedish dilemma" involves the question of how to deal with nuclear power and the implications of the nuclear power decision on energy, environmental, and economic policy.

The dilemma arises because Sweden must now decide how it will deal with the 1980 referendum on nuclear power. Nuclear power has been controversial in Sweden since the first plant was constructed in 1972. Critics focused primarily on safety issues, and the question was brought to a head in a national referendum in 1980 that determined that Swedish nuclear power should be phased out. As a result of the referendum, the Swedish Parliament took two steps. The Parliament decided that no further nuclear power reactors would be licensed. In addition, it decided that the existing nuclear reactors should not be allowed to operate beyond the expected lifetime of the youngest reactor, often taken to be the year 2010.

Much has happened since the nuclear referendum. The debate faded away, was then revived by the Chernobyl accident, and, subsequently, was again swept under the political rug. In the last few years, however, it became apparent that the time for temporizing had passed and that the issue would soon need to be addressed. Although the deadline of 2010 might seem to be far into the future, in terms of planning large-scale energy systems, a decade and a half is but the blink of an eye. In addition, with the rise of concerns about global warming, the impact of closing the nuclear power plants on carbon dioxide emissions must be addressed. Most recently, Sweden has decided to deregulate its electricity generation industry, raising questions about the interaction of nuclear power and deregulation. In the end, all sides of the Swedish nuclear debate recognize that orderly planning requires settling the issues soon so that long-term investment decisions inside and outside the energy sector can be planned and undertaken.

The dilemma is: how should Sweden deal with the 1980 referendum? Should Parliament phase out existing nuclear power facilities in accordance with the 1980 referendum? Or, should the earlier decisions be reconsidered in light of changing priorities and evidence? How should we incorporate the fact that Swedish nuclear power has been relatively low cost and safe, and that developing a new capacity will probably be much more expensive? What are the economic costs and benefits of a nuclear phaseout? What are the environmental impacts of a shutdown? How do Sweden's commitments to combat global warming fit into its nuclear commitment? And, is the nuclear discontinuation an expensive luxury, appropriate in a period of robust economic growth and an expanding state, that must now be rethought along with other important political decisions and programs of the last half century?

THE SCOPE OF THE STUDY

The purpose of this study is to analyze the Swedish nuclear debate. But the implications are much broader, for they allow us to consider the trade-offs between two of the most important environmental issues of today—concerns with the hazards of nuclear power and those involving future climate change. Much as Sweden was a crucible for testing the policies of the welfare state, just so is Sweden grappling with the central environmental dilemmas today. The approach developed here can be applied equally well to other countries or regions as they think through their own environmental dilemmas.

In considering the nuclear dilemma, the focus will be primarily economic. At the same time, the major environmental, health, political, and safety issues must be given adequate consideration. The philosophy of this study is that Sweden should choose its energy options in a way that will maximize the sustainable consumption possibilities of its residents. In stating this objective, it should be emphasized that "consumption" should be broadly interpreted to include nonmarket as well as market elements, including appropriately valued environmental, health, and safety aspects. Due consideration must also be given to international considerations, including the value of international environmental goals and a free and open trading system, as well as the value of a stable set of rules and property rights inside Sweden.

Within that general framework, the study that follows will examine the economic costs and benefits of different energy options. It seems likely that electricity demand will be growing in the years to come, so a nuclear phaseout will require finding replacement electricity. Replacement power will come primarily from domestic production, possibly supplemented with electricity imports. One set of questions involves the economic impact of replacing nuclear power with the next best alternative. What will be the impact on real national income? What are the implications for health and environmental quality in Sweden if nuclear power is replaced by coal, oil, natural gas, or renewable sources? What are the impacts on electricity prices under different options? Will this be inflationary, and will the higher prices injure Swedish competitiveness in international trade? These questions will be addressed below.

Another key set of issues involves the interaction between the nuclear power dilemma and the proposed deregulation of the electricity industry in Sweden and more broadly in Europe. What will happen to electricity prices under deregulation? Will nuclear power become more or less viable in a deregulated environment? Will the plants become "stranded assets" in Sweden as has occurred in Britain and is likely to occur in the United States? Is a nuclear phaseout more or less costly in a deregulated environment?

Part I of this study explores the structure of the Swedish energy system, the environmental issues involved in the nuclear debate, issues of climate-change policy, and the regulatory environment. In Part II, I turn to a detailed analysis of the options available to Sweden. The emphasis there will be to use state-of-the-art economic and energy modeling to estimate the costs and benefits of different approaches. At the same time, we must recognize that models are not truth machines and cannot resolve fundamental political or philosophical conflicts. They provide a menu of choices, along with the price tags, but it is up to individuals and governments ultimately to decide what to do.

Part II begins with an outline of the outlook for growth in electricity demand over the next two decades. I then show how supply reductions such as a nuclear phaseout would affect the electricity market. The subsequent sections describe an empirical model of the Swedish electricity system that has been constructed for this study as a way of estimating the costs and benefits of different courses of action. This model, which is called the Swedish Energy and Environmental Policy (SEEP) model is described verbally. The model is then applied to the nuclear phaseout debate, estimating the costs and benefits of different schedules, with and without considering other environmental constraints, particularly that concerned with carbon dioxide emissions.

The single chapter of Part III summarizes the Swedish situation by recapitulating the overall economic and political consequences of Sweden's pursuit of its nuclear power options. The issues of internal and international political commitments as well as fiscal, health, safety, and environmental impacts are considered in the context of Sweden's having reached a turning point in its postwar history.

Before plunging into the study, a word is in order about the techniques that follow. This study relies in part on mathematical and statistical models of the Swedish energy system. Such an approach is taken without apology. There is simply no way to think systematically and rigorously about the interaction of the many forces that interact in the economy and the energy system without analytical and numerical economic models of the kind that are employed here. At the same time, I will try to explain the logic of the analysis and the shortcomings of the models so that the reader can judge where the results seem reliable and where they are highly conjectural. In the end, of course, the decision about how to proceed must be taken by people on the basis of human values and human judgments about likely future outcomes. But, the outcomes are more likely to be satisfying ones if the judgments are well informed rather than based on fuzzy and contradictory hunches.

ENDNOTES

[1]See Gunnar Myrdal, *An American Dilemma: The Negro Problem and Modern Democracy,* New York, London, Harper & Brothers, 1944.

2

Historical Background

This chapter explores the background of the Swedish energy system. It begins with a brief sketch of the Swedish economy for those whose contact with Sweden comes mainly from the movies or a brief visit. It next examines the historical origins of energy use in Sweden, focusing primarily on electric power; maps out the growth in demand; and describes the major sources of supply. The chapter then compares the Swedish energy system with those in other major industrial countries, looking at the level and trends in energy-output and electricity-output ratios, examining comparative electricity prices, and comparing the Swedish electricity supply system with those of other countries.

A BRIEF SKETCH OF THE SWEDISH ECONOMY

Sweden is a medium-sized open economy with a large natural resource base for its small population. Sweden has a population of 8.5 million with a per capita gross domestic product (GDP) (at market exchange rates) of $25,000. Sweden is increasingly integrated into the European economy, having joined the European Union in 1993. Table 1 shows the relation of Sweden to the total twelve-country version of the European Union. Clearly, Sweden is well endowed with land, forests, Nobelity, and human capital.

The Swedish economy was one of the European "miracle" states from around 1870 until the first oil shock in 1973. After that time, Sweden came under increasing strains, with the most recent shock being the tight German monetary policy after German reunification in 1990 and the Scandinavian recession that ensued upon

Table 1. Sweden's Position in the European Union.

Item	*Swedish share of European total*[a] *(percent)*
Population	2.5
Land area	18.9
GDP	3.5
Forest land	52.6
Energy production	5.0
Carbon dioxide emissions	2.1
Nobel laureates in economics	20.0
Patents awarded in U.S.	6.0

[a]Europe is 12 country total.

Source: Swedish Ministry of Finance, *The Medium Term Survey of the Swedish Economy*, Stockholm, 1995.

the economic collapse of the Soviet Union and Eastern Europe. The strains of the early 1990s were exacerbated by the ill-fated decision of Sweden to peg the krona against the European Community basket of currencies. This decision led to a sharp rise in interest rates and deepening recession in 1992.

Since 1990, output has been stagnant, unemployment has soared, and government debts and deficits have increased sharply. Real GDP growth, which had averaged 4% from 1950 to 1970, declined to 2% from 1980 to 1990, and fell at an average annual rate of 1.7% from 1990 to 1993. Open unemployment climbed from under 2% in the 1960s and 1970s to over 8% in 1993. The government deficit rose to 15% of GDP. Over the period from 1970 to 1991, Swedish per capita GDP (on a purchasing-power basis) fell from third to fourteenth among the countries of the Organisation for Economic Co-Operation and Development (OECD). As a result, real household incomes have fallen and Sweden has been forced to rethink many of its most cherished institutions, including the cradle-to-grave welfare state that inspired much thinking and emulation in other high-income democracies.[1]

One of the major constraints on Swedish economic policy is its high degree of openness. In 1990, Sweden's export-output ratio was 30% for all GDP and 45% for industrial production; moreover, those ratios are increasing sharply with the integration of Sweden into the European and world markets. Sweden's exports come heavily

Table 2. Major Sources of Energy Supply in Sweden, 1970 and 1994.

	TWh electric or thermal[a]		*TWh thermal*[b]	
	1970	*1994*	*1970*	*1994*
Petroleum	350	204	350	204
Nuclear power	0	73	0	219
Hydroelectric power	41	59	123	177
Coal	18	28	18	28
Natural gas	0	9	0	9
Other	48	88	48	88
Total	457	461	539	725
Energy/GDP (1970 = 100)	100.0	68.7	100.0	91.7

[a] "TWh" or "terawatt hour" represents one trillion (10^{12}) watt hours.

[b] Hydroelectric and nuclear energy converted to thermal energy at a ratio of 3 to 1.

Source: NUTEK, *Energy in Sweden: Facts and Figures, 1995.*

from resource sectors, with 55% of exports composed of such industries as forest production and iron and steel. Many of these resource-intensive exports are also energy intensive, which is threatening to these industries if electricity prices rise sharply over the coming years. At present, Sweden conducts very little trade in energy (other than petroleum products), although that may change if the European electricity market is deregulated.

The Swedish energy system is dominated by petroleum, nuclear power, and hydroelectric power; see Table 2 for a description of the major sources of supply. The first two columns show the conventional measure of energy content, while the second two measure in terms of thermal equivalent energy. Sweden has become much more energy efficient by the first measure, but much less so by the second measure, because of the substitution away from fossil fuels and toward electricity generated from nuclear and hydroelectric power.

THE SWEDISH ELECTRICITY SYSTEM

Historical Origins

The electric power industry in Sweden began with power plants for town lighting in the 1870s and was first fueled with oil and coal;

exploitation of hydroelectric power soon followed.[2] The industry began to take its current organizational form when the Swedish electricity public enterprise, Vattenfall ("waterfall"), was created in 1909. Waterpower proved to be a low-cost source of power, particularly when Vattenfall could borrow at the low government bond rate, and hydroelectric power dominated the growth of supply until the 1960s.

A central feature of the Swedish electricity system is the geographical separation of production and consumption for a substantial fraction of electricity supply. Most hydroelectric power is generated in the sparsely populated and heavily rivered north, while most of the population clusters around in the fjords and archipelagoes of the south. The distances involved—as much as one thousand miles—implied that building an efficient long-distance transmission system was crucial. The responsibility for planning and operating the national grid was, after 1946, largely entrusted to the state-owned enterprise Vattenfall. The different firms were coordinated in an informal "club" rather than through the formal regulation seen in countries like the United States.

By the 1950s, it became clear that hydroelectric power could not meet the growing demand envisaged for the rapidly growing Swedish economy. Well before the oil shock of the 1970s, Sweden determined that it would be unwise to rely on insecure foreign oil supplies. In 1956, the government approved an approach known as the "Swedish line." This policy determined that nuclear power using heavy-water reactors—for use in both electricity and heat production—would be the next stage of Swedish power. This decision was designed to keep the entire nuclear fuel cycle within Sweden.[3]

Today, Sweden has a diversified electricity industry with a large number of generators and distributors, but for the most part the members of the industry function as vertically integrated monopolists. The major electricity producers (as of 1994) are shown in Table 3.

The striking feature of the Swedish energy system is the high degree of concentration. The largest producer, Vattenfall, has more than half of production and the top four have 85% of the market. This fact is of substantial importance for the market structure under deregulated conditions. Vattenfall was until 1992 a state agency under the Ministry of Industry, at which point it was transformed into a wholly state-owned corporation, Vattenfall AB, and directed

Table 3. Major Electricity Producers in Sweden, 1994.

Company	*Total production (tWh)*	*Nuclear power (tWh)*
Vattenfall	72.9	42.9
Sydkraft	25.7	15.5
Gullspång	8.4	4.0
Stockholm Energi	9.6	5.1
Stora Kraft	5.5	1.7
Others	15.6	1.0
Total	137.7	70.2

Note: One tWh is a terawatt-hour or one billion kilowatt-hourw (kWh).

Source: The Swedish Power Association.

to behave in a commercial manner. There was discussion about privatizing Vattenfall, but that proposal has been postponed.

Another important reform in 1992 was the separation of the operation of the long-distance grid from Vattenfall. At that time, a separate state-administered agency, Svenska Kraftnät ("Swedish Power Network"), was established to operate the long-distance grid. This separation was strategically crucial. In a modern electricity system, the long-distance grid is the single most important source of national monopoly power.[4] By separating the transmission system from the other components, Sweden established the conditions for an effective and competitive deregulated internal market.

Electricity Use and the Growth of Demand

Electricity use in Sweden grew rapidly over most of the period since World War II, as it did in most other industrial countries. From 1970 to 1993, electricity consumption in Sweden grew at an average rate of 3.5% per year and reached over 140 terawatt-hours (or billions of kilowatt-hours) in 1993. A distinct slowdown occurred after 1984, however, with total use growing at an average annual rate of only 0.3% per year over the period 1987 to 1993.

The major components of electricity demand are the residential and commercial (R&C) sector and the industrial sector. Electricity use in the R&C sector has grown most rapidly, at an average rate of 5.2% per year over the 1970–1993 period, and comprises about half of total electricity use. Sweden has a relatively high use of electricity for heating purposes compared with other industrial countries. Sweden has favored electric heating because of the lack of natural gas as well as the low electricity price compared with the prices of other fuels.

Historically, industrial use of electricity has also been relatively large in Sweden, particularly in paper and pulp and in iron and steel. In 1993, industrial use comprised 35% of the country's total electricity consumption. This sector's consumption has grown relatively slowly; it has actually declined since 1990, largely because of the economic crisis. Industrial output fell by an average of 2% annually over the 1990–1993 period, while electricity consumption fell at an average rate of 2.8% per year over this period. Other sectors—transportation and district heating—are relatively small electricity uses.[5]

Major Sources of Supply

Most of Sweden's electricity today is produced by either hydroelectric power or nuclear power, with the two being roughly equal in importance. In addition, less than 10% of electricity is produced using higher marginal-cost sources, such as combined heat and power, oil-fired plants, and gas turbines. Renewable energy has fired much popular imagination but little electricity, with solar, tide, and wind presently constituting only 0.03% of generation. Table 4 shows the major sources of electricity generation in Sweden in 1973 and 1993.

The History of Nuclear Power in Sweden

The first major nuclear initiative in Sweden came shortly after the 1955 Geneva conference, "Atoms for Peace." The policy known as the "Swedish line" was designed to secure a high degree of self-sufficiency in energy supply for the long term. With no indigenous energy sources other than hydroelectric power, and with that source evidently limited, there was no available source other than petroleum. A government study noted that the "risk of disturbances [to the oil market] due to war or risk of war is manifest and is magnified by the Middle East...."[6] There was a remarkable clairvoyance in these predictions. Concerns about the impact of an oil market

Table 4. Major Sources of Electricity Generation in Sweden, 1973 and 1993.

Source	*1973 (tWh)*	*1993 (tWh)*
Coal	0.50	3.02
Oil	15.18	3.08
Gas	n	0.92
Nuclear	2.11	61.40
Geothermal	n	n
Solar, tide, wind[a]	n	0.05
Comb. renew. & waste	0.40	2.14
Hydroelectric	59.87	74.60
Total[b]	78.06	145.20

[a]Includes wave, ocean, and other.

[b]Electricity generated equals gross production minus amount of electricity produced in pumped storage plants.

n = less than 0.01 tWh.

Source: IEA Statistics, *Electricity Information 1994*, IEA/OECD 1995.

disruption was a driving force in sending Sweden down the nuclear path.

The first steps along the path proved a fiasco. The technocratic planners of the 1950s and 1960s chose to design a heavy-water reactor using the abundant but low-grade Swedish ores. The result was a heavy-water plant in Marviken, built on a Swedish design, with Swedish labor, capital, engineering, and natural uranium. The reactor proved expensive, technical problems accumulated, and a test run led to fears of an accident—all of which led to a 1970 decision to close the reactor before it became operational. Tragedy turned to farce when the owners decided to fuel the plant with oil, which led critics to declare that Marviken was "the world's only oil-fueled nuclear power station."

The first operational nuclear-fueled station in Sweden was in Ågesta, a suburb of Stockholm. This small (sixty-megawatt [MW] thermal) plant was one of the few in the world designed primarily to produce heat rather than electricity. It cost seven times the original estimate, produced heat for ten years, and was then closed in 1974 just when the rise in oil prices might have made production profitable. Because of cost overruns, this approach was also abandoned.

In the mid-1960s, Swedish designers (particularly Allmanna Svenska Elektriska Aktiebolaget [ASEA], which was a technologically sophisticated power equipment producer) turned to what became the predominant nuclear power technology, the light-water reactor. In 1965, ASEA proposed to produce a full-scale plant in Oskarshamn.

After the debacles of Marviken and Ågesta, particularly after Swedish designers adopted the mainstream reactor designs and undertook collaborative efforts with U.S. companies, Sweden's nuclear experience proved more successful. The next generation of plants was built largely on time and on budget. Indeed, it is ironic that the nuclear power industry that is proposed to be shut down is one of the major nuclear power success stories of the world. The first of Sweden's twelve operating nuclear power plants was commissioned in 1972, and the final two reactors were finished in 1985. Swedish power plants are of two basic configurations of the light-water reactor. Nine of the twelve plants are boiling-water reactor systems designed by ABB-ATOM, while the three remaining plants are pressurized-water reactor systems designed by Westinghouse.

Table 5 shows the basic parameters of the different systems. In addition, this table shows the availability of the reactors during 1994. By the mid-1990s, the oldest plants were showing some signs of aging. There were significant shutdowns for safety reasons, and, as can be seen in Table 5, one of the twelve plants was shut down for the entire year.

Those who have mainly followed the sad history of the American nuclear power industry may be surprised to learn of the performance of post-1970 Swedish systems. They have on the whole been built on schedule and without major cost overruns. They have provided reliable service, with high levels of availability, and at production costs that are well below the costs of alternative sources of supply. A comparison of nuclear with hydroelectric costs performed by Hjalmarsson indicates that, even using the relatively low electricity prices in Sweden, the present values of the last two 1100-MW nuclear power plants have positive present value and have earlier present-value break-even points than do the most recent hydroelectric plants.[7] I will return later to a more detailed examination of the performance of Swedish nuclear power, but the short descrip-

Table 5. Operational Swedish Nuclear Power Plants, 1993.

Power plant	*Reactor type*	*Electric output (mWe)*		*Availability 1994 (percent)*	*Commercial operation (year)*
		net	*gross*		
Barsebäck 1	BWR	600	615	86.7	1975
Barsebäck 2	BWR	600	615	77.2	1977
Forsmark 1	BWR	968	1006	91.6	1981
Forsmark 2	BWR	969	1006	92.6	1981
Forsmark 3	BWR	1158	1197	93.5	1985
Oskarshamn 1	BWR	445	462	0.0	1972
Oskarshamn 2	BWR	605	630	88.7	1975
Oskarshamn 3	BWR	1160	1205	89.0	1985
Ringhals 1	BWR	795	825	78.3	1976
Ringhals 2	PWR	875	905	84.5	1975
Ringhals 3	PWR	915	960	91.7	1981
Ringhals 4	PWR	915	960	84.8	1983

Notes: "Availability" refers to the fraction of the year that the plant was available for electricity generation. PWR refers to pressurized-water reactors while BWR refers to boiling-water reactors.

Source: Kärnkraftsäkerhet och Utbildning AB, *Operating Experience in Swedish Nuclear Power Plants, 1994,* Stockholm, 1995.

tion is that, from an economic and technical point of view, after the initial fiasco with the heavy-water reactor and the heat plant, the Swedish nuclear power industry has been remarkably successful.

International Trade in Electricity

The Scandinavian countries are moderately integrated in their electricity systems, and the potential for fruitful trade in electricity and other energy products is one of the major issues that must be addressed in Sweden and Europe over the coming years. Unlike most industrial commodities, which are increasingly subject to free or close-to-free trade, electricity has been subjected to high barriers to trade and entry in most of Europe. The major barriers are high levels of state ownership and regional monopolies in most of the European markets. These institutional barriers are reinforced by relatively small international transmission linkages—which are controlled by government agencies—that serve as powerful barriers to international trade.

To see the extent of autarchy in the electricity market, we can examine the Nordel countries (Norway, Sweden, Finland, Denmark, and Iceland). In 1990, total gross trade (the average of imports plus exports) constituted 8% of generation among the Nordel countries. Given their geography and technical systems, Norway and Sweden form a natural economic unit for purposes of electricity generation, transmission, and pooling. Yet, most electricity is generated and consumed in each country. Sweden's net exports today constitute less than 2% of its production.

To determine the extent to which barriers have affected trade, we can compare the heavily protected national markets in Europe with the largely open markets in the United States. One comparison would be to examine the "trade" of the six New England states. This region is slightly more populous and slightly smaller than Sweden, but it has a completely integrated electricity market operated as a power pool. (A power pool allocates electricity production to minimize the cost of production among all the plants in the region.) In 1992, the ratio of gross trade to production for the six New England states was 33% compared with 2% for Sweden. Another example is California, which imports 44% of its electricity consumption from other regions. These examples suggest that the barriers to trade in the European electricity market have significantly reduced international trade in electricity.

One of the major questions that arises for the coming decades is the impact of deregulation and potential privatization on international trade in electricity. As we will see below, electricity prices to end users are currently much lower in Sweden and Norway than in the northern European continental market, especially in Germany. Electricity prices are maintained at elevated levels in Germany partially because of the high degree of integrated regional monopoly and partially because of the enormous subsidies to the coal industry, which are passed on to electricity consumers. It is therefore unclear what the net result on Sweden would be of complete liberalization of electricity markets in northern Europe. It might lead to a more rapid phasing out of coal subsidies and lowering of wholesale prices in northern Europe; or it might lead to significant exports from Scandinavia. In the longer run, however, it seems most likely that the price level with free electricity markets in

Europe would settle close to the marginal cost of replacement power from fossil or nuclear fuels. We will see that this outcome parallels the conclusion that we draw below about the long-run price level for Sweden itself.

In summary, it seems likely that barriers to international trade in electricity will decline in the coming years, although Europe is currently far from a deregulated market. The prospect of importing electricity from other countries to replace nuclear power is one that must be considered. As we will see below, however, there seems little prospect that a substantial fraction of the nuclear generation can be imported at prices that are significantly below the cost of alternative sources of domestic supply.

THE SWEDISH ENERGY SYSTEM IN INTERNATIONAL CONTEXT

How does the Swedish energy system compare with those of other major industrial countries? The answer to this question is helpful for our understanding of the strengths and weaknesses of Swedish energy policy and for informed judgments about how a nuclear phaseout might affect the Swedish economy. In this section we compare the levels and trends of energy-output ratios, sources of supply, and prices in Sweden with those in other major countries. These comparisons are useful because energy technologies and fuels are widely available and can be purchased in the open market.

Exchange-Rate Conventions

Many of the national data for this study were denominated in Swedish kronor. However, because the language of the marketplace and many of the international comparisons are stated in U.S. dollars, we have provided most estimates for this study in terms of 1995 U.S. dollars. We provide a brief description and rationale in this section.

International price comparisons are difficult for a number of reasons. One reason is that countries have differing tax regimes, including different sales, value-added, and direct taxes, and these are difficult to untangle. Even trickier is the question of which exchange rate to use. We have generally chosen market exchange rates for the comparisons in this study. The reason for this choice is that energy fuels and equipment are traded and tend to have equalized prices on world markets (subject to trade restrictions). Moreover, if European markets are to be deregulated and interconnected, the appropriate prices for determining comparative advantage and trade flows are prices at market exchange rates. While purchasing-power parity (PPP) exchange rates may be more appropriate for purposes of living standards, market exchange rates are appropriate for analyzing market decisions in the energy market.

A further difficulty arises from the fact that the Swedish krona has been quite unstable over the last decade. The krona-dollar exchange rate averaged 6.7 kronor to the dollar over the 1986–1995 period, with a high of 7.8 and a low of 5.8. Over the period 1991–1995, which is the period most relevant for the present study, the market exchange rate averaged 7 kronor to the dollar. This is the value we chose for exchange rate conversions in the present study.

For comparative purposes, we can ask the impact of using PPP exchange rates, which are the common standard in comparing international prices. For 1990, Sweden's PPP exchange rate is calculated to be approximately 7.6 kronor to the dollar.[8] Thus, like Japan and much of Europe, Sweden's market exchange rate is overvalued relative to its purchasing power. If PPP prices had been used instead of market exchange rates, prices in Europe and Japan would generally be lower than those shown later in Table 10. For example, if the PPP dollar-yen exchange rate were 150 rather than 100, then the electricity price in PPP units would be approximately two-thirds of the level shown in the figures and tables. For Sweden, the extent of overstatement is small because the deviation of the market from the PPP exchange rate is small in the 1991–1995 period.

Energy-GDP Ratios

Sweden is often lauded as the model energy-efficient society. Yet, a careful look at the data indicates that Sweden in fact is highly energy intensive. Figure 1 shows the history of energy consumption, electricity consumption, and the ratios of energy and electricity to GDP since 1973. As that figure shows, the energy-GDP ratio has declined modestly over the last quarter century in Sweden. To a certain extent, this decline is an accounting fiction as costlier but

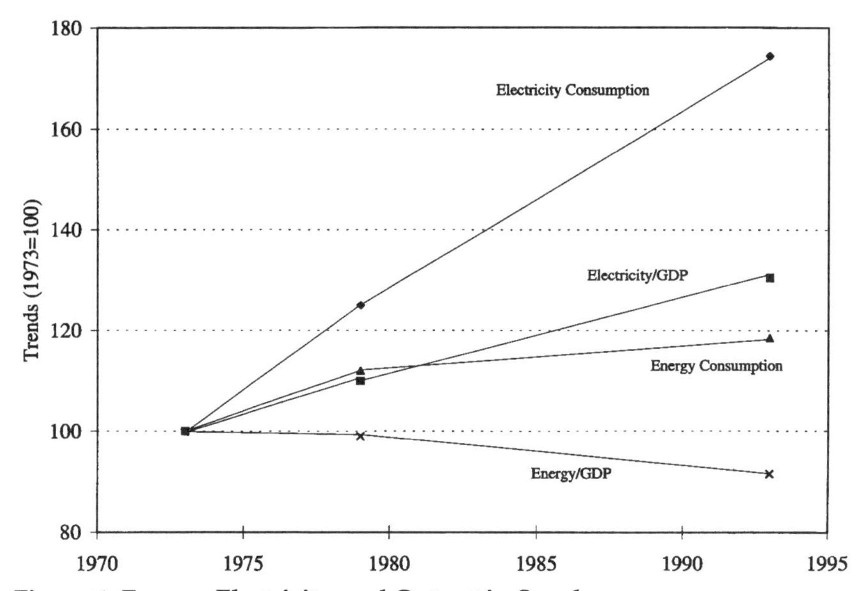

Figure 1. Energy, Electricity, and Output in Sweden.

more energy-efficient sources replaced less costly but less energy-efficient sources (this was shown in Table 2). But the overall trend is definitely toward higher energy efficiency. By contrast, the electricity-GDP ratio rose over this period.

Comparative data on energy, electricity, and output by country are shown in Tables 6 and 7 for major countries and for the OECD as a whole. The trends are shown in Figures 2, 3, and 4. Figure 2 shows the levels and trends in total primary energy consumption per unit of GDP. For these calculations, we use GDP measured at official exchange rates to reflect the fact that energy fuels and technologies are largely tradable goods. Figure 2 shows that Sweden is in the middle of the pack in terms of the energy intensity of its economy. The most energy intensive economy is Canada, while the least energy intensive is Japan. One interesting feature of Figure 2 is that Sweden is virtually unique among the major countries in showing stability in its energy intensity. Both OECD Europe and the entire OECD had declines in their energy-output ratio of 25% over the period. By contrast, Sweden's energy-GDP ratio declined only 8% over the same period.

The trends in electricity consumption per unit of GDP are shown in Figure 3. Sweden had by far the largest "electrification"

Table 6. Total Primary Energy Supply, Electricity Consumption, and Electricity Generation by Country.

	TPES (Mtoe)[a]			*Electricity consumption (tWh)*			*Electricity generation (tWh)*[b]		
	1973	*1979*	*1993*	*1973*	*1980*	*1993*	*1973*	*1979*	*1993*
Canada	161.0	190.9	220.7	223.2	307.5	451.4	270.0	359.1	527.28
France	176.8	190.3	233.7	160.0	231.6	356.1	182.5	240.4	467.82
Germany	338.1	372.4	337.2	337.6	419.2	467.1	374.3	467.5	521.93
Japan	324.0	355.2	457.4	421.6	520.2	804.6	465.3	585.3	899.05
Norway	15.1	18.8	22.2	61.0	75.1	102.4	73.0	88.8	119.57
Sweden	39.3	43.9	47.1	69.4	86.1	122.9	78.0	95.9	145.2
U.K.	220.9	219.8	216.9	242.4	243.3	295.1	281.3	298.6	321.59
U.S.	1723.2	1869.7	2028.6	1715.9	2099.7	2963.9	1965.5	2359.6	3389.93
OECD Europe	1221.5	1330.8	1440.8	1280.2	1613.7	2179.8	1462.9	1812.9	2469.04
OECD total	3550.8	3911.8	4389.6	3738.2	4695.2	6674.7	4281.1	5287.5	7608.82

[a]TPES = total primary energy supply; Mtoe = million tons oil equivalent.

[b]Electricity generated equals gross production minus amount of electricity produced in pumped storage plants.

Source: IEA Statistics, *Electricity Information, 1994*, OECD/IEA, 1995.

Table 7. TPES, Electricity Consumption, and Electricity Generation as a Share of GDP.

	TPES/GDP			*Electricity consumption/GDP*			*Electricity generation/GDP*		
	1973	*1979*	*1993*	*1973*	*1980*	*1993*	*1973*	*1979*	*1993*
Canada	0.49	0.45	0.38	0.68	0.72	0.79	0.82	0.85	0.92
France	0.22	0.20	0.19	0.20	0.24	0.30	0.23	0.25	0.39
Germany	0.30	0.29	0.20	0.30	0.32	0.27	0.34	0.36	0.30
Japan	0.21	0.19	0.15	0.28	0.27	0.26	0.30	0.31	0.29
Norway	0.25	0.24	0.19	1.02	0.91	0.90	1.22	1.12	1.04
Sweden	0.24	0.24	0.22	0.42	0.46	0.56	0.47	0.51	0.67
U.K.	0.31	0.29	0.22	0.35	0.32	0.30	0.40	0.39	0.33
U.S.	0.47	0.44	0.35	0.47	0.50	0.51	0.54	0.56	0.59
OECD Europe	0.25	0.24	0.20	0.27	0.28	0.30	0.30	0.32	0.34
OECD total	0.33	0.31	0.25	0.35	0.37	0.38	0.40	0.42	0.44

Note: GDP is measured in billions of U.S. dollars, 1990 prices.

Source: IEA Statistics, *Electricity Information, 1994*, OECD/IEA, 1995; and National Accounts 1960–1992, OECD, 1994.

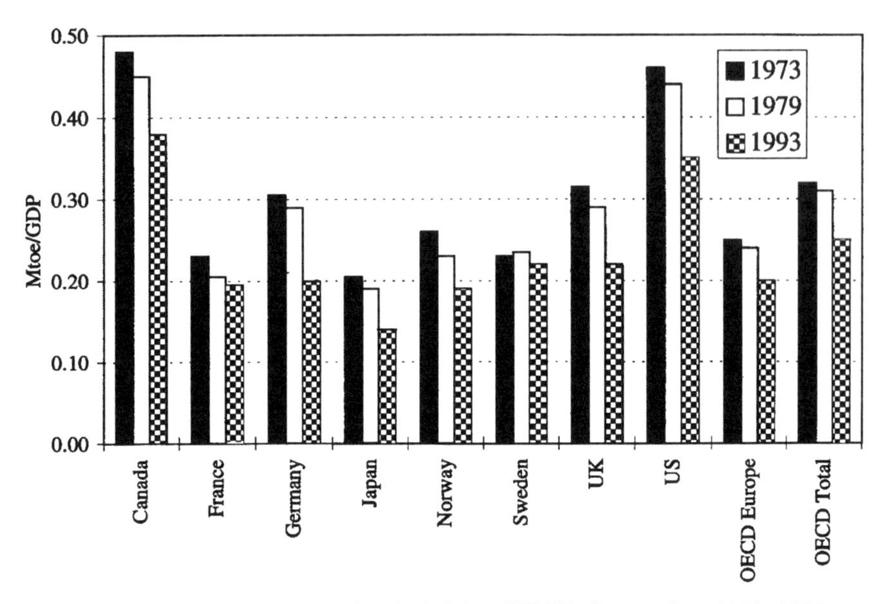

Figure 2. Energy-Output Ratios in Major OECD Countries, 1973–1993.

Note: Mtoe/GDP indicates millions of tons of oil equivalent per billion dollars of GDP in 1990 prices.

Source: International Energy Agency, *Electricity Information 1994*, IEA/OECD, 1995.

over the 1973–1993 period. France and Canada, also countries with low electricity prices, showed a significant but less marked increase in electrification, while electricity-output ratios in most other countries were stable or declined. By the end of the period, 1993, the electricity-output ratios in Sweden, Canada, and Norway were far ahead of other countries.

Figure 4 shows the ratio of generation (as opposed to consumption) to total output. These trends are quite close to those of consumption because of the modest level of international trade in electricity that we discussed above.

Sources of Supply

How does the Swedish energy supply system compare with thoses in other major industrial countries? Table 8 shows the sources of generation for major countries in 1993. A number of features stand out in these data. First, Sweden has a much higher share of both

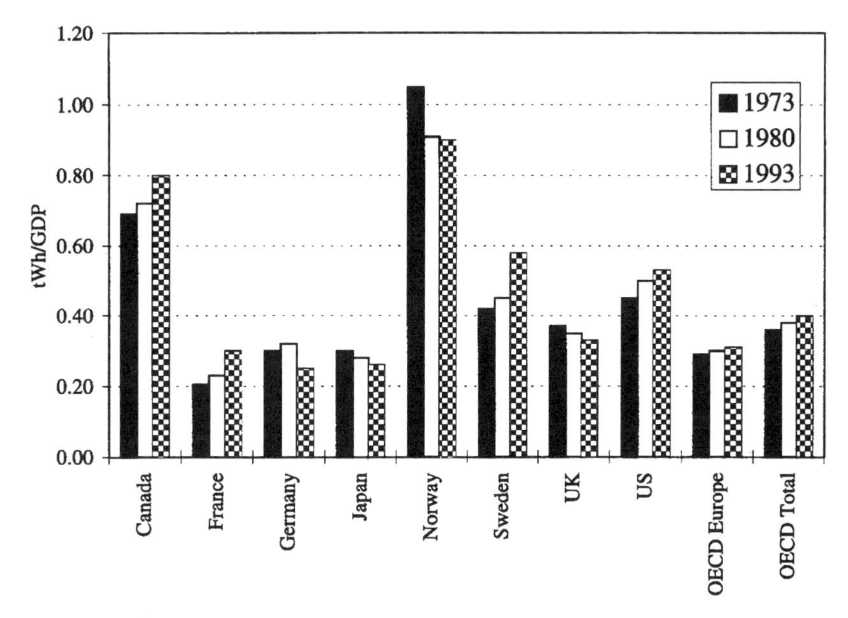

Figure 3. Electricity Consumption-Output Ratios in Major OECD Countries, 1973–1993.

Note: tWh/GDP indicates terawatt-hours of electricity per billion dollars of GDP in 1990 prices.

Source: International Energy Agency, *Electricity Information 1994*, IEA/OECD, 1995.

nuclear and hydroelectric generation than do most countries. Only France and Belgium have a larger share of their electricity generated from nuclear power than does Sweden. Among the hydroelectric countries, Norway is most dependent on hydroelectric power, and Sweden has a similar share to Switzerland and Canada.

In addition, Sweden is remarkable in its lack of reliance on conventional fossil fuels for electricity generation. The OECD as a whole generates approximately 60% of its electric power from fossil fuels, whereas that figure is close to 6% for Sweden. That figure will almost surely change considerably (directly or indirectly) if nuclear power is phased out.

One hardy perennial in the garden of ideas for energy technologies is the question of the role of renewable resources such as wind, solar, and geothermal. It is instructive to note that for the OECD as a whole, even two decades after the first oil shock, only 0.5% of electricity is generated using renewable fuels.

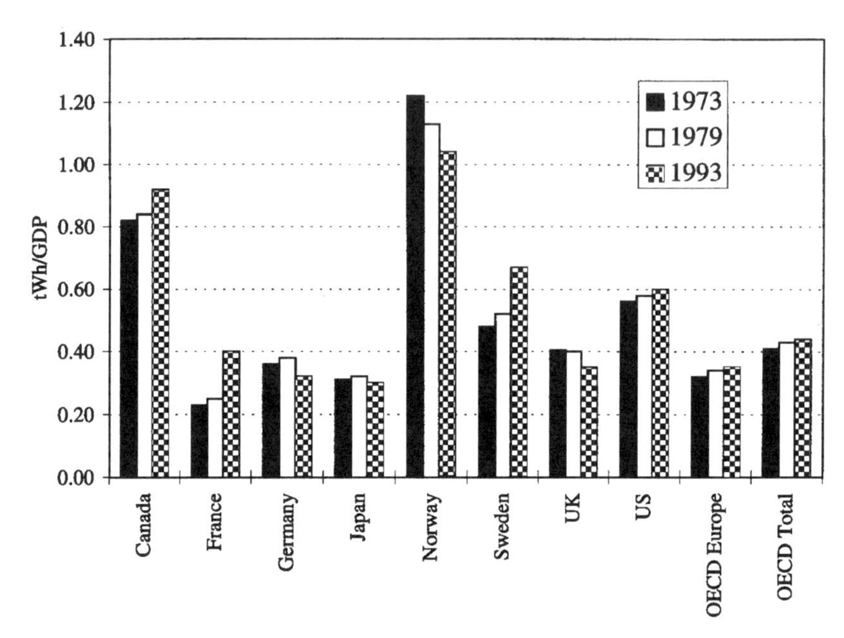

Figure 4. Electricity Generation-Output Ratios in Major OECD Countries, 1973–1993.

Note: tWh/GDP indicates terawatt-hours of electricity per billion dollars of GDP in 1990 prices.

Source: International Energy Agency, *Electricity Information 1994,* IEA/OECD, 1995.

Price Comparisons

Figure 5 shows the trend in electricity prices in Sweden over the last twenty-five years. Unlike most industrial countries, electricity prices have been remarkably stable. Industrial electricity prices are (in deflated terms) almost exactly what they were before the oil shocks of the 1970s, while residential prices have risen primarily because of increases in taxes.

How do electricity prices in Sweden compare with those in other countries? Tables 9 and 10 provide the basic data for international comparisons, with Table 9 showing the data in national currencies and Table 10 translating these into U.S. dollars. Table 10 and Figure 6 provide the most useful comparisons because they examine industrial prices excluding taxes, which are indicative of the wholesale cost of electricity (including distribution).

Table 8. Gross Electricity Production by Fuel and Country, 1993 (percent of total production).

					Combustible fuels[b]		
	Nuclear	*Hydro*	*Geothermal*	*Solar/Wind*[a]	*Electricity plants*	*CHP plants*	*Total (tWh)*
Australia	–	10.10	n	n	86.19	3.71	163.80
Belgium	59.18	1.44	n	0.01	29.15	10.22	70.80
Canada	17.98	61.38	n	0.02	20.63	n	527.40
Denmark	–	0.08	n	3.05	10.17	86.70	33.70
Finland	32.52	22.23	n	n	12.79	32.47	61.20
France	78.01	14.38	n	0.12	7.49	n	472.00
Germany	29.19	4.08	n	0.02	66.70	n	525.70
Italy	–	19.97	1.65	0.16	67.74	10.49	222.80
Japan	27.49	11.63	0.20	0.00	60.68	n	906.70
Netherlands	5.13	0.12	n	0.23	n	94.52	77.00
Norway	–	99.59	n	0.00	0.31	0.09	120.00
Sweden	42.06	51.64	n	0.03	0.26	6.01	146.00
Switzerland	38.24	59.96	n	0.00	n	1.80	61.10
U.K.	27.66	1.76	n	0.07	66.12	4.39	323.00
U.S.	18.97	8.88	0.52	0.12	61.67	9.84	3411.30
OECD Europe	32.86	20.13	0.16	0.10	39.36	7.39	2488.10
OECD total	23.68	16.97	0.41	0.09	51.99	6.87	7657.50

[a]Includes tide, wave, ocean and other.

[b]Total production from electricity and CHP (combined heat and power) plants using coal and coal products, oil and oil products, gas and combustible renewables and waste.

n = less than 0.01 tWh.

Source: IEA Statistics, *Electricity Information, 1994*, OECD/IEA, 1995.

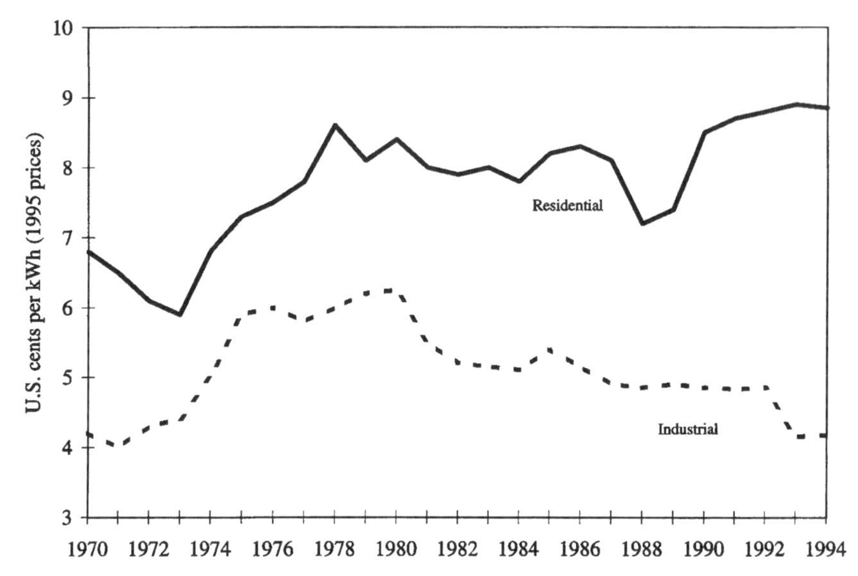

Figure 5. Electricity Prices in Sweden, 1970–1994.

Table 9. Prices of Electricity by End User, 1993 (in national currency per kWh).

	Industry		*Households*	
	Incl. tax	*Excl. tax*	*Incl. tax*	*Excl. tax*
Canada	0.0506	0.0506	0.0808	0.0808
Finland	0.2768	0.2598	0.4605	0.3605
France	0.3097	0.3082	0.8289	0.6599
Germany	0.1477	0.1374	0.2794	0.2260
Italy[a]	0.1454	0.1273	0.2295	0.1928
Japan[b]	0.0186	0.0076	0.0264	0.0249
Sweden	0.2750	0.2750	0.6382	0.4252
U.K.	0.0451	0.0451	0.0754	0.0754
U.S.	0.0486	0.0486	0.0834	0.0834

[a]1000 Italian lire/kWh

[b]1000 Japanese yen/kWh

Source: IEA Statistics, *Electricity Information, 1994*, OECD/IEA, 1995.

Table 10. Prices of Electricity by Country for Different End Users, 1993.

	Price[a]				
	Industry		*Households*		*Exchange rate (national currency per $US)*
	Incl. tax	*Excl. tax*	*Incl. tax*	*Excl. tax*	
Canada	0.0392	0.0392	0.0626	0.0626	1.29
Finland	0.0484	0.0454	0.0805	0.0630	5.72
France	0.0547	0.0545	0.1464	0.1166	5.66
Germany	0.0895	0.0833	0.1693	0.1370	1.65
Italy[b]	0.0925	0.0810	0.1460	0.1226	1.57
Japan[c]	0.1673	0.0683	0.2374	0.2239	0.11
Sweden	0.0353	0.0353	0.0819	0.0546	7.79
U.K.	0.0673	0.0673	0.1125	0.1125	0.67
U.S.	0.0486	0.0486	0.0834	0.0834	1.00

[a]Prices given in U.S. dollars per kWh; dollar converted at market exchange rates for 1993 shown in last column.

[b]1000 Italian lire/kWh.

[c]1000 Japanese yen/kWh.

Source: International Energy Agency, *Electricity Information, 1994*, OECD/IEA, Paris, 1995.

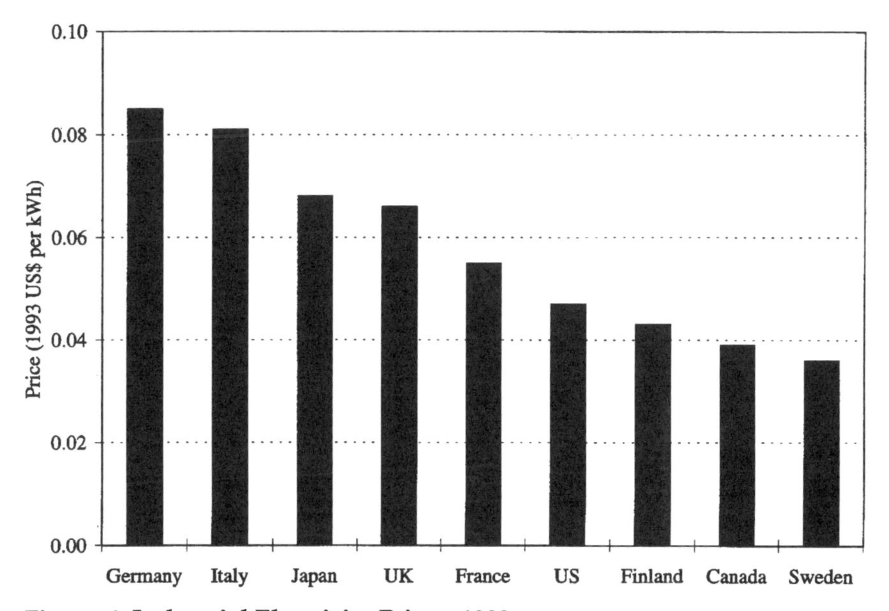

Figure 6. Industrial Electricity Prices, 1993.

Electricity prices vary widely across countries, with Germany having the highest of the prices and Sweden the lowest. The pretax industrial price of electricity in Sweden, at 3.5 U.S. cents per kilowatt-hour (kWh), is below most other countries and far lower than the prices in Germany (at around 8.3 cents per kWh) or Italy (at 8.1 cents per kWh). Pretax prices for households and small users in Sweden are again much lower than in major industrial countries and are approached only by the low prices found in Canada.

The tax treatment of electricity varies widely across countries. Industrial electricity is generally taxed lightly or not at all, perhaps in accordance with standard public-finance precepts that intermediate goods should not be taxed. Sweden, in fact, had among the heaviest taxation of electricity in 1991, but since that time Sweden has abandoned taxation of industrial electricity altogether.

The general impression given by these data suggests the following conclusions. Because of its favorable geography (plentiful potential for hydroelectric power) and successful exploitation of nuclear power, Sweden has a large supply of relatively inexpensive electricity. This has led to low electricity prices, with the consequent high ratio of electricity to output, and a concentration on highly energy-intense industries. Given the low and falling energy prices, particularly those of electricity, it is not surprising that Sweden's electricity-output ratio has been rising sharply, a phenomenon that is unique among major industrial countries.

Sweden has, however, come to the end of this path. There is little hydroelectric power left to harness, and policymakers have decided to foreclose further hydroelectric development. The nuclear power option is at stake with the possibility of shutting down half of the low-cost Swedish electricity supply. There are no obvious low-cost sources of electricity at hand. This, in a nutshell, is the Swedish nuclear power dilemma.

ENDNOTES

[1]The data and analysis in this section come largely from Swedish Ministry of Finance, *The Medium Term Survey of the Swedish Economy*, Stockholm, 1995; and Assar Lindbeck et al., *Turning Sweden Around*, MIT Press, Cambridge, Mass., 1995.

[2]This historical survey is based on Lennart Hjalmarsson, "From Club Regulation to Market Competition in the Scandinavian Electricity Supply Industry," in Richard Gilbert and Edward Kahn (eds.), *International Comparisons of Electricity Regulation*, New York, Cambridge University Press, 1996.

[3]For a description of the background of the "Swedish line," see Stefan Lindström, "The Brave Music of a Distant Drum," *Energy Policy*, July 1992, pp. 623–631.

[4]On the importance of the transmission grid for effective competition, see for example William Hogan, "Contract Networks for Electric Power Transmission," *Journal of Regulatory Economics*, vol. 4, no. 3, 1992, pp. 211–242.

[5]Many of the data from this section are drawn from NUTEK (the Swedish National Board for Industrial and Technical Development), *Energy in Sweden, 1994*, Stockholm, August 1994 and *Energiläget I siffror*, NUTEK, Stockholm, June 1995.

[6]Report of the Swedish government from 1956. This quote and the discussion of the decisions in the early period of Swedish nuclear power are from Stefan Lindström, "The Brave Music of a Distant Drum," *Energy Policy*, July 1992, pp. 623–631.

[7]See Lennart Hjalmarsson, "From Club Regulation to Market: Competition in the Scandinavian Electricity Supply Industry," in Richard Gilbert and Edward Kahn (eds.), *International Comparisons of Electricity Regulation*, New York, Cambridge University Press, 1966, p. 54.

[8]This value is calculated from World Bank, *World Development Report, 1993*, Washington, D.C., Oxford University Press, 1993.

3

Issues To Be Considered

We have up to now described the key elements of the Swedish energy system—the context in which the decision and debate about the Swedish nuclear phaseout take place. We turn next to a discussion of the key issues in the decision. Most of the issues involve questions of environmental policy. At the center is the future of the nuclear option in Sweden and elsewhere, particularly the safety and cost of nuclear power. What risks are posed by continued operation of nuclear power—both routine operating risks and risks of catastrophic accidents? What is the safety record of Swedish nuclear power plants? How did these questions arise in the original nuclear referendum and the debate surrounding that referendum? And what are the lessons of historical experience over the fifteen years since the original Swedish referendum?

In asking these questions we must be ever mindful that the nuclear power decision does not take place in a vacuum. Clearly, a phaseout of nuclear power will have major economic consequences for Sweden, and in subsequent sections we will estimate the economic consequences. But there are environmental consequences as well. For example, if nuclear power is phased out, some of the electricity production will be replaced by fossil fuel–powered electricity; this will lead to higher carbon dioxide (CO_2) emissions. The higher CO_2 emissions will threaten to violate another environmental policy in Sweden, the commitment not to increase its CO_2 emissions above 1990 levels. If the replacement electricity were generated by fossil fuels, Sweden would need to worry about potential increases in conventional pollutants like the oxides of sulfur and nitrogen with the consequent economic and health damages from that source; coal would be the dirtiest and natural gas the cleanest of the fossil fuels, but none comes without some environmental complications. Another replacement option would be to increase hydroelectric power, but this would butt into another self-imposed constraint, that of leaving the remaining undammed northern rivers untouched. And, as we will see below, other fuel cycles have their own health and safety consequences.

In short, there is no costless solution, no option without significant economic and environmental—and even perhaps psychological—consequences. In this section, we discuss the major options to be considered in this study.

THE NUCLEAR DEBATE AND THE NUCLEAR REFERENDUM

The 1980 Swedish referendum on nuclear power had its roots in a political struggle that involved energy policy. Nuclear power has been controversial since its inception. To some extent, this was inevitable because it was billed as the "peaceful use of atomic energy," thereby reminding people of the awesome destructiveness of nuclear fission. Even today, people routinely confuse power-generating nuclear reactors with nuclear weapons.

The origins of the nuclear referendum grew from a debate between the Social Democratic Party and the Center (formerly the Agrarian) Party. This debate led to a tentative compromise in early 1979 that would limit Swedish nuclear power to the twelve stations then under construction (and currently operating). At this point, the Three Mile Island accident intervened by casting doubts on both the reliability of nuclear power and, perhaps even more important, damaging the credibility of the nuclear power industry. At this point, the Social Democrats withdrew their support for nuclear power and called for a national referendum. Given the deep fissures through Swedish parties and public opinion, a national referendum was seen as the only way of resolving the conflict.[1]

It should be noted that in Sweden referenda are rare events, having been used but four times since World War I. The other three involved prohibition, establishment of a national retirement system, and a change to driving on the right side of the road. Unlike the plague of Californian referenda, they are advisory rather than binding on Parliament. Of the three previous referenda, Parliament followed the popular will on two, while it eventually disregarded the overwhelming desire of Swedes to continue to drive on the left side of the road.

The 1980 referendum had three alternatives. All three alternatives endorsed the view that there should be no further construction of nuclear power after the twelve plants in operation or under construction had been completed. The first two alternatives held that these twelve nuclear power plants should be used for their "technically safe lifespan," but no further nuclear power plants should be constructed. The third alternative held that the six operating reactors should be phased out within ten years (that is, by 1990) and that the six reactors under construction should not be allowed to operate. The results of the referendum were the following:

- Alternative 1 (phase out at end of plants' lifetimes): 18.9%
- Alternative 2 (phase out at end of plants' lifetimes with provisos): 39.1%
- Alternative 3 (stop construction and phase out nuclear power by 1990): 38.7%
- Blank ballots: 3.3%

The total vote constituted 76% of the eligible electorate.

The referendum can be interpreted as saying that 60% of those expressing a view opted for use of the twelve reactors but no construction of further reactors. Both of the first two alternatives held that "nuclear power shall be phased out at a rate compatible with electrical power requirements for the maintenance of employment and national well-being. ... Safety issues are the determinants of the order in which the reactors will be shut down." In Alternative 2, preferred by the plurality, an explanation of the text stated that "In anticipation of other, less risky types of energy, we want to use nuclear power by having the twelve reactors in operation no longer than their technically safe lifespan of about 25 years."

The referendum was then implemented by a parliamentary decision that set the phaseout date at 2010. This date was apparently selected on the assumption of a 25-year "technically safe lifespan" of the nuclear power plants, although there may have been some confusion between the accounting lifetime and the useful economic lifetime (an important point to which we return below).

THE INTERPARTY AGREEMENT OF 1991

After the referendum and the parliamentary deliberations of the early 1980s, governments temporized, assuming that the phaseout would not become a live issue until the first nuclear plants reached the projected end of their operating lifetimes sometime toward the end of the 1990s. In the early 1990s, discussions were held, and an important political milestone was passed with the *Interparty Agreement on Energy Policy in Sweden* of January 15, 1991. This agreement—by the Social Democrats, the Liberal Party, and the Center Party—covered many points, but the key elements were not to reopen the commitment to the nuclear phaseout and to add certain other commitments on environmental policy. This agreement set the ground rules for the reconsideration of the nuclear referendum in 1995.

The Text of the Agreement

The major points of the Interparty Agreement follow.

On general policy objectives: "The aim of energy policy is to secure the long-term and short-term supply of electricity and other energy on internationally competitive terms, thereby furthering good economic and social development in Sweden. Energy policy must be based on what is naturally and environmentally sustainable. ... The use and development of all energy technology must conform to strict requirements concerning safety and care of the environment. ... Secure supplies of electricity, reasonably priced, are an important prerequisite of the international competitive strength of Swedish industry."

On the nuclear phaseout: "In 1980 ...the Riksdag [the Swedish Parliament] declared that nuclear power was to be phased out at a rate compatible with electrical power requirements for the maintenance of employment and national well-being. At the same time, the Riksdag recommended a commitment to closing down Sweden's last reactor not later than 2010. ..."

The juncture at which the phase-out of nuclear power can begin and the rate at which it can proceed will hinge on the results of electricity conservation measures, the supply of electricity from environmentally acceptable power production and the possibilities of maintaining internationally competitive electricity prices.

On hydropower. "The unharnessed rivers and the river stretches which the Riksdag has excluded from hydropower development will remain protected in future. Accordingly, the expansion of hydropower will not be allowed to exceed the hydropower plan adopted by the Riksdag."

On climate-change policies: "Carbon dioxide emissions resulting from the combustion of fossil fuels are having an impact on climate. This being so, it is essential that the combustion of fossil fuels should wherever possible be avoided."

On deregulation: "One can predict an increasing amalgamation of Europe's national electricity and natural gas systems into a multinational energy market. ... Development of this kind deserves to be viewed in a positive light. ... A European electricity market can eventually lead to an equalization of electricity prices in the countries concerned."

Reading between the Lines

What do these statements of principle signify? As in all policy pronouncements, they cover as much as they reveal and contain inconsistent objectives. Nonetheless, the following would seem to be the major features of the agreement:The general statement of principles is extremely broad, but it is possible to discern three general objectives: First, electricity and energy policy must be consistent with sound economic policy. Low energy and electricity prices—and ones that reflect marginal costs of production including environmental costs—contribute to high and growing living standards. Second, environmental objectives must be appropriately included in energy-policy decisions. Some of the objectives are purely national (such as the hydroelectric limitation) while others are global public goods (such as slowing harmful climate change), but they must be appropriately addressed. There is a strong inclination toward the use of taxes as regulatory instruments where they can meet the environmental goals. Third, the agreement is sensitive to the fact that Sweden is a small open economy and must tread carefully when taking steps that might harm the international competitiveness of its key industries. Because of the low energy prices in Sweden (as discussed above), Sweden has a special niche in world trade because its exports tend to be energy intensive, natural resource intensive, and skill intensive. Raising electricity prices too sharply would clearly hurt Swedish exports (a point to which we return below), and attention must be paid to this impact of energy policy.

The discussion of nuclear power in the Interparty Agreement broke no new ground and resolved none of the outstanding issues. In its discussion of the Interparty Agreement in the Summary of the Government Bill (1990/91:88) On Energy Policy, the government stated, "The question as to when nuclear power is to be phased out has not formed the subject of renewed assessment or decision during the discussions between the parties." This could alternatively read, "When it is possible on controversial issues, temporize."

The clear statement on the development of further hydroelectric power suggests that expansion of hydroelectric power has very little political support in Sweden. This is consistent with my conversations with key figures in Sweden, in which there was virtually no support for expanding hydroelectric power.

The discussion of climate-change policy in the Interparty Agreement was in fact a muddle. The Interparty Agreement noted that in 1988 the Riksdag had endorsed a statement, "As a national sub-target, it should be stated that carbon dioxide emissions should not be increased above their present-day level." It went on to note that this objective had already been breached and that the target "will probably be hard to achieve within the next few years." The Interparty Agreement then endorsed moving to a European target (in contrast to the national subtarget just endorsed) that "should be for total carbon dioxide emissions in the year 2000 in the countries concerned not to exceed the present level and to decline thereafter." This suggests that there should be a lid on total European emissions not on the emissions of Sweden or individual countries.

The Interparty Agreement gave its blessing to deregulation of electricity markets, although there was virtually no discussion of the topic. The one strange twist in the discussion is the statement that a deregulated market would lead to electricity-price equalization across countries, which seems to contradict the general statement of principles that Sweden should attempt to maintain the competitiveness of its export industries, who benefit from relatively low electricity prices.

CLIMATE-CHANGE POLICY

Since the Interparty Agreement of 1991, two further major policy developments in Sweden and elsewhere involve climate-change policy and electricity deregulation. These two developments are next described. It should be emphasized that each of these is an enormous area, so the summary below can give only a broad overview of the questions involved. We begin with a discussion of climate change.

The Scientific Background

One of the major developments affecting the future of nuclear power has been the growing concern with greenhouse warming. What is meant by greenhouse warming or the greenhouse effect? The greenhouse effect is the process by which radiatively active gases like carbon dioxide (CO_2) selectively absorb radiation at different points of the spectrum and thereby warm the surface of the earth. The greenhouse gases (GHGs) are transparent to incoming solar radiation but absorb significant amounts of outgoing radiation. There is no debate about the importance of the greenhouse effect, without which the earth's climate would resemble the moon's.[2]

Concern about the greenhouse effect arises because human activities are currently raising atmospheric concentrations of greenhouse gases. The major GHGs are carbon dioxide (emitted primarily from the combustion of fossil fuels), methane, and chlorofluorocarbons (CFCs). Scientific monitoring has firmly established the buildup of the major GHGs over the last century.

While the historical record is well established, there is great uncertainty about the future course of climate change. On the basis of climate models, scientists project that a doubling of the atmospheric concentrations of CO_2 will in equilibrium lead to a one- to five-degree (Celsius) warming of the earth's surface; other projected effects include an increase in precipitation and evaporation, a small rise in sea level over the next century, and the potential for hotter and drier weather in midcontinental regions.

To translate these equilibrium results into a projection of future climate change requires scenarios for emissions and concentrations. Using rudimentary emissions modeling, the Intergovernmental Panel on Climate Change (IPCC), an international panel of distinguished scientists led by the Swedish scientist Bert Bolin, projected that "business as usual" would produce a three- to six-degree (Celsius) warming in 2100. Current integrated assessment models (which often incorporate elaborate economic and energy models along with simplified climate modules) suggest an increase in global mean temperature in 2100 that is two to three degrees (Celsius) warmer than our current climate.[3] While there is no consensus among different approaches, virtually all projections are worrisome because climate appears to be heading out of the historical range of temperatures witnessed during the entire span of human civilizations.

What are the likely impacts of climate change on human civilizations and nonhuman systems? All research indicates that climate change is likely to have highly variable effects on different sectors and different countries.[4] In general, those sectors that depend most heavily on unmanaged ecosystems—that is, are heavily dependent upon naturally occurring rainfall, runoff, or temperatures—will be most sensitive to climate change. Agriculture, forestry, outdoor recreation, river systems, and coastal activities fall in this category. Because developing countries tend to have a larger fraction of their economies devoted to agriculture, especially rainfed agriculture, this means that they are more subject to the vagaries of climate change.

By contrast, most economic activity in high-income countries has little *direct* interaction with climate. For example, cardiovascular surgery and microprocessor fabrication are undertaken in carefully controlled environments and are unlikely to be directly affected by climate change. Studies indicate that as much as 85% of gross domestic product in high-income countries will be largely unaffected by climate change over the next century.

Recent concerns about climate change have focused on the potential for major surprises and on the impacts on ecosystems. While warming may seem benign, it has major and unpredictable impacts on weather patterns, ocean currents, sea-level rise, river runoffs, storm and monsoonal tracks, desertification, and other geophysical phenomena. Many scientists and ecologists view these changes and uncertainties with alarm. Recent results of ice-core analysis have heightened concerns about future warming. The Greenland Ice-Core Project (GRIP) has found evidence that very rapid temperature changes in the Greenland area have occurred in warm periods (earlier results had found climate instability associated only with transitions between warm and glacial periods).[5] Recent scientific evidence suggests that the climate may have experienced multiple different locally stable equilibriums, perhaps because of changing ocean circulation patterns. A recent survey indicates that many climatologists believe that there is a significant probability that a doubling of atmospheric concentrations of CO_2 will move the earth's climate system to a different stable climatic equilibrium.[6] All these recent findings have increased the level of concern about inducing major changes in our sensitive climate system.

The burden of the evidence points increasingly to the tenuousness of our understanding of the relationships among climate change, human societies, and ecosystems. If the mainstream scientific view proves correct, substantial global warming is inevitable over the next century even if stringent control measures are taken; the only question is whether the extent of global temperature rise will be modest (say in the one- to two-degree Celsius range over a century) or large (say more than three degrees Celsius over a century). Pasteur observed that chance favors the prepared mind, yet nations are still largely unprepared to deal with the possible future impacts of climate change.

International Negotiations on Climate Change

Over the last decade, the call to arms against climate change has attracted increasing attention among policymakers, culminating in the United Nations Conference on Environment and Development in Rio de Janeiro (the "Rio Summit") in June 1992. At the Rio Summit, nations agreed on the Framework Convention on Climate Change, which held that nations should achieve "a stabilization of greenhouse gas concentrations in the atmosphere at a level that would prevent dangerous anthropogenic interference with the climate system" (Article 2). In addition, Annex 1 countries (the Organization for Economic Cooperation and Development, eastern

Europe, and the former Soviet Union) agreed to adopt national policies that would "individually or jointly" stabilize anthropogenic GHG emissions at 1990 levels by the year 2000. Developing countries are given a grace period and simply pledge to provide information about their GHG emissions.

The Rio Summit was supplemented by the Berlin Conference of the Parties in 1995. The Berlin Conference strengthened the Rio commitments by agreeing that Annex 1 countries will "set quantified limitation and reduction objectives within specified time frames, such as 2005, 2010, and 2020, for their anthropogenic emissions. ..." At present, international deliberations concerning climate change are in ferment. Generally, governments have expressed great concern about the prospects of climate change, but few governments have taken any costly steps to curb their GHG emissions. The U.S. government is a signal example of the tension between good intentions and good politics. During the 1992 campaign, Vice President Al Gore stated, "The task of saving the Earth's environment must and will become the central organizing principle of the post-cold-war world." After an initial attempt to levy energy taxes in 1993, the Clinton-Gore administration proposed no major initiatives on global warming. Then, in 1996, the United States floated a proposal that appeared to endorse a global emissions target, although there were no accompanying proposals to implement this target at the national level and the United States has to date been unwilling to take the most modest steps to control its GHG emissions. Similarly, in Europe environmental ministries in many governments proposed a high European carbon tax (equivalent to over $100 per ton of coal). This proposal fell prey to national political squabbling and the skepticism of finance ministries in most countries. The world has far to go in implementing the commitments made at the Rio Summit.

Greenhouse Gas Emissions and Climate Policy in Sweden

Sweden has been an active participant in the deliberations on climate change. The pathbreaking work on the greenhouse effect originated with a Swedish chemist, Svante Arrhenius. In the modern era, a Swedish atmospheric scientist, Dr. Bert Bolin, has been the active and outspoken leader of the Intergovernmental Panel on Climate Change, the group that has provided the scientific and policy analysis that has supported the Framework Convention on Climate Change as well as the follow-up meeting.

Sweden's policies and analysis concerning climate change have been described in its report under the Framework Convention.[7] The basic policy as hammered out in the Interparty Agreement and the 1991 government bill on energy is described above. The more concrete goals were set forth in the government bill, *Actions to Counteract Climate Change* (1993). This bill stated that Sweden shall, in accordance with the climate convention, stabilize its emissions of carbon dioxide at the 1990 level by the year 2000. After that date, emissions shall be reduced. The strategy covers all greenhouse gases, but pending new knowledge no targets are set for greenhouse gases other than carbon dioxide. In calculating its emissions, current practice is to omit biofuels (under the theory that they are simply recycling carbon), exclude the contribution of forests (as too uncertain), and exclude other greenhouse gases. In addition, it is customary to include only direct carbon dioxide emissions (which would occur through burning coal to generate electricity) but to exclude indirect emissions (such as those that would come from importing electricity that is generated by burning coal).

An important question involves whether Sweden might claim credit for reductions in carbon dioxide emissions that are financed by Sweden but take place in other countries. Such emissions-trading practices are called "joint implementation." While it is recognized that it will be cost-effective to reduce carbon dioxide emissions in other countries rather than in Sweden, this possibility has not been included in Sweden's national action plan.

Sweden's "counted" carbon dioxide emissions[8] for 1990 were 61.3 million metric tons. This represents a significant decline from around 100 million tons in 1970. The decrease occurred largely as a result of replacing coal-fired plants with hydroelectric and nuclear power stations. Table 1 shows the breakdown of carbon dioxide emissions for the year 1990.

Notwithstanding Sweden's policies and pronouncements, Sweden's national action plan foresees a modest growth in carbon dioxide emissions over the next decade. *Sweden's National Report* projects that emissions will rise from 61.3 million tons of CO_2 in 1990 to 63.8 million tons in 2000 and 67.8 million tons in 2005. These projections do not include any increase in emissions that would occur in the event of a nuclear phaseout. Projections by NUTEK made for

Table 1. CO_2 Emissions in Sweden, 1990.

	Thousands of metric tons
Not counted[a]	
Biofuels	21,737
Forestry	–34,368
Counted[b]	61,256
Nonenergy	6,081
Energy	55,175
Nonelectric	51,015
Transportation	23,092
Other	27,923
Electric[c]	4,160

[a]"Not counted" are those emissions or absorptions which are not conventionally counted as part of Sweden's contribution to CO_2 emissions.

[b]"Counted" are those emissions which are conventionally counted as part of Sweden's contribution to CO_2 emissions.

[c]This includes a credit for that part of combined heat and power that is devoted to electricity production.

Source: Swedish Ministry of the Environment and Natural Resources, *Sweden's National Report Under the United Nations Framework Convention on Climate Change*, September 1994, Stockholm.

the Energy Commission envision the possibility of a major increase in CO_2 emissions, with some of the scenarios containing a near doubling of emissions by 2020.[9]

We will consider below the relationship between Sweden's climate-change policy and its nuclear power decision. Even with the current nuclear power production, it will clearly be difficult for Sweden to keep its commitments to stabilize carbon dioxide emissions. If Sweden chooses to phase out its nuclear power plants, meeting that commitment will go from difficult to near impossible.

DEREGULATING THE ELECTRICITY MARKET

International Trends

The trend toward deregulating electricity markets has been under way for two decades, and Sweden comes relatively late—but well structured—to participate in this movement. I will first describe the recent international developments concerning deregulation of electricity markets. I then analyze the Swedish deregulation plan and the interaction of deregulation with the nuclear phaseout.

Historically, electricity has been one of the most heavily regulated industries. Many countries have regulated electricity through government ownership of major facilities, while others have used traditional rate-of-return public utility regulation to control the prices and markets of electric utilities. The pure government-ownership model is seen today in France, where Electricité de France is in essence a government-controlled, vertically and horizontally integrated monopoly. At the other extreme is the United States, with a predominance of state-regulated, vertically integrated local monopolies supplying relatively large areas. The Swedish model is a hybrid, often described as a "club system,"[10] in which a government-owned supplier (Vattenfall) tends to act as a price leader, and operations are coordinated through informal agreements.

Electricity has been heavily regulated because it was typically cast as a "natural monopoly," which is an industry where a single firm can provide a specific good or service at a lower cost than can multiple firms. In analyzing the electricity market, it is useful to divide the electricity industry into three parts: generation, transmission, and distribution. Up until recently, it was commonly assumed that because of the high transmission costs and significant economies of scale in generation, the entire industry should be treated as a natural monopoly. In the last decade, this view has been challenged, and only the transmission and the small-customer distribution network are now viewed as having major elements of natural monopoly. The shift has come in part as a result of the successful breakup of vertically and horizontally integrated telecommunications networks and partly because of the evident exhaustion of the economies of scale in generation. The common view among independent specialists today is that, while the generation segment is technically complex, the optimal scale of plant is sufficiently small so that reasonably competitive markets can emerge among alternative generators.

These theoretical presumptions have been strengthened by experience around the world in the actual deregulation of the electricity markets. The most important and relevant experiments for Sweden are the British and the Norwegian cases. The former, which has been carefully studied and is highly influential because of its pathbreaking nature, will be briefly described here.[11] Before the reforms, the British system (strictly speaking, that of England and Wales) consisted of two horizontally integrated state-owned segments—one for generation and one for transmission. The British government decided to break up and privatize the industry. The generation segment was divided into two generating companies, one of which (National Power) was to contain the nuclear power stations and 60% of the conventional power, while the other (PowerGen) was to contain the remaining plants. The transmission segment was transferred to the National Grid Company, which was to be a regulated monopoly. The distribution or supply segment was divided into twelve regional electricity companies.

One of the most radical features of the British reforms was the establishment of a "spot market" to coordinate generation. Many systems coordinate electricity generation by a centralized "pool" system, whereby a central dispatcher brings power stations on line in increasing order of their marginal production costs. Under the British spot market, each generator must declare its offers of quantities and prices of supply for each half-hour of the following day. The National Grid Company then ranks these offers on the basis of costs, and, together with estimates of demand, the National Grid then sets up a plan for least-cost operations for the coming day. This also generates a series of spot prices, which have proven highly variable (as is appropriate for an industry in which the marginal cost varies so greatly depending upon the load).

No conclusive evaluation of the British reforms is possible at this time, but a few tentative conclusions can be reached on the basis of evidence to date. First, the reforms have led to a system that not only functions but also appears to allocate electricity smoothly in a spot market. Second, by opening up the generation market and separating generation from transmission, the British reforms have induced significant entry of generation, with an extra entry of combined-cycle gas turbines of eleven gigawatts (one million kilowatts) planned by 1996 compared with a total of about sixty gigawatts. Third, the planned sale of the nuclear stations was unsuccessful, as there proved to be no market for them, so the government was forced to retain them and charge a special fee levied on fossil-fuel generation to subsidize the nuclear power generation. Fourth, and perhaps most relevant for the Swedish plan, there has been significant criticism of the creation of a duopoly generating industry, in which the two firms clearly have incentives and market power to raise prices above the competitive level. Finally, it appears that wholesale and retail electricity prices in Britain have not fallen appreciably since the reforms.

Britain was one of the earliest countries to liberalize its electricity market, but in the last few years a number of other countries have taken steps to deregulate certain parts of the market. Although the features of the deregulation differ greatly across countries given the different initial institutional conditions, the major common themes in the different countries are the following.[12]

- In the generation stage, most countries are moving toward a system with competition among generators. Sometimes this is achieved by forcing open access to transmission systems, while in other cases generation and transmission are being separated so that the incentives to exclude potential competitors are reduced.
- In the transmission stage, it is increasingly being recognized that the transmission system is the key to competition in the electricity industry. Where systems are vertically integrated (as in many systems in the United States and in the individual countries of Europe), lack of access to transmission systems is a formidable barrier to entry. Moreover, as we noted earlier, where national governments require permitting and approval of international linkages, this forms a major barrier to international trade in electricity. The forced separation of generation from transmission in Britain was probably the most effective means of ensuring competition; by contrast, the efforts to enforce open access have yielded meager results.
- The transmission and retail distribution stages generally continue to have price and service regulation in most countries. The major innovation in these stages is to move away from traditional rate-of-return regulation to "RPI minus *X*" regulation. Under traditional rate-of-return regulation, prices are determined as the cost of production plus an authorized rate of return on invested capital. This approach has very weak incentives to economize and is biased toward capital-intensive production techniques. A radical alternative approach is to have prices set as the inflation rate of the retail price index (RPI), less a normative efficiency improvement (*X*). This new approach

mimics a competitive market in that the regulated firm becomes a price taker, so that any cost reduction flows directly into profits. The major disadvantage of this approach is uncertainty about the appropriate X rate, which should represent the target rate of cost reduction relative to the economy as a whole. This new approach has been introduced in the British electricity reforms, but the regulators were extremely conservative in choosing $X = 0$, implicitly assuming that there would be no differential productivity improvement in the electricity industry. If X is set incorrectly for too long a period, the system either will degenerate into an unprofitable industry or will allow excess profits. The British regulators appear to have been more averse to losses than to excess profits, although we can expect that adjustments will occur in the coming years.

The Swedish Deregulation Proposal

Sweden has considered a deregulation of its electricity industry over the last decade. The Swedish debate received impetus from the British experience discussed earlier, from deregulation of the Norwegian market in January 1991, and from discussions during the 1990s about a single market for electricity in the European Union. We first describe the proposal and then present a brief evaluation.

The Swedish Proposal. The Swedish deregulation proposal was first put forth by the conservative government, and the succeeding left-center coalition government, after some initial hesitation, decided to proceed with the plan pretty much as outlined by the conservative government. The basic principles of the Swedish proposal as of early 1996 are the following:[13]

- The central principle behind the reform of the electricity market is to establish a clear separation between the production and sale of electricity on the one hand and the transmission of electricity on the other. The production and sale of electricity should be subject to competition, while transmission should be regulated and monitored to ensure efficient operation.
- The transmission network will be divided into two parts: the national transmission grid and the regional and local grids. The national grid will be owned and operated by Svenska Kraftnät,

which is a government agency, while the regional and local grids will be owned by power companies while remaining subject to government regulation.

- Svenska Kraftnät will be responsible for ensuring service reliability and balancing supply and demand in the short run. Svenska Kraftnät will also be responsible for creating a market for electricity. It is not clear whether this market will be a pool-type arrangement or a true spot market like the British example, although recent discussions suggest that a joint Swedish-Norwegian spot market may be established.
- Owners of transmission lines must provide open third-party access (TPA). That is, they will be obliged to connect the lines and plants of other companies to their own networks and to transmit electricity on reasonable terms. It is unclear at this time how transmission pricing will be regulated.
- Foreign trade in electricity will continue to be regulated by the government. In particular, permits will be required to build and to operate international transmission linkages. There will therefore continue to be significant barriers to international trade.

Evaluation. As with most major changes in complex systems, it is difficult to predict the outcome of a change as thorough as that proposed for the Swedish electricity system. A few summary comments will suffice for the purpose of this study.

First, Sweden has, by a happy accident of history, found itself with an industrial structure that is quite well designed for an efficient deregulation of its electricity industry. The major advantage of the current system is that the primary element of natural monopoly—the long-distance transmission grid—is already largely unified as an independent entity and can therefore serve as a vehicle for competition in the wholesale trade in power. This system contrasts with those in Germany and the United States, where the presence of vertically integrated monopolists is a major barrier to entry for competitors. In these two countries, as with vertically integrated monopolies like the former Bell telephone system, it will be extremely difficult to mandate TPA when such access is against the economic interests of the owners of the transmission system. In other words, the open access to the Swedish national grid that results from the separation of ownership of generation and supply from transmission will be extremely useful in promoting effective competition in electricity generation.

Second, the potential for effective international competition is still unclear. The extent to which foreign producers can serve as effective competitors is limited by the transmission links between Sweden and other countries. At present, the total capacity of the international linkages between Sweden and other countries is 34 terawatt-hours (tWh) out of a total generation of 145 tWh. In effect, the maximum presence of foreign producers on the Swedish market today is approximately 25% of capacity. This figure overstates the extent of potential foreign competition, however, because many of these links are not available at a particular time and total trade is far below capacity. To be effective competitors in the Swedish market, the international linkages would have to be significantly larger.

Third, the major issue raised by the Swedish proposal involves the current dominant producer, Vattenfall. As noted above, Vattenfall has a share of almost one-half of the electricity market. The key question for Sweden is whether to allow Vattenfall to remain a dominant firm in the generation and distribution market. In the old system, where Vattenfall was a regulated government enterprise, the main potential abuse was exploitation of its excessively low cost of capital and overcapitalization of plant.[14] In the new economic environment, Vattenfall will be a dominant domestic producer and will have significant incentives to understate available capacity and thereby raise prices—whether this be through the medium of a power pool or through the impersonal operation of a spot market. The danger of allowing a dominant producer has been carefully analyzed in the British case, where a dominant generator was allowed to prevail. Experts on that system have argued that this has led to gaming of the spot market and monopolistic abuses. Indeed, electricity prices have actually risen after the British deregulation, which is difficult to explain except as exercise of market power.

Concerns about the role of Vattenfall in a deregulated market were raised in a study by Lars Bergman and Bo Andersson. They find that a deregulation keeping the current market structure would lead to a significant increase in prices. They further conclude that "if the electricity market consists of at least five electricity-producing firms of equal size, their possibility to influence the market price is reduced significantly."[15] How could the Bergman-Andersson target of a structure in which no firm has more than 20% of the generating capacity be implemented? This could be met by breaking Vattenfall into two parts and ensuring that foreign suppliers had transmission linkages that could provide net imports into Sweden of at least one-fourth of the domestic Swedish generating capacity.

Defenders of Vattenfall's current status make two points. First, they note that a better analogy would be Norway, where the dominant producer (Statkraft) does not appear to have exercised its market power. However, the question of whether some dominant producers have refrained from exercising their monopoly power is not persuasive; it is better to ensure that the market structure is one in which no company *has the ability* to raise prices rather than to rely on the good citizenship of companies. A second point in defense of the current structure is that Vattenfall is a medium-sized duck on a small pond, but with the integration of the European market it will become a medium-sized duck on a huge lake. In other words, it cannot be a major player in a wider Nordic or European market. Indeed, the argument goes, in the larger market Sweden may become better served if it has a large company like Vattenfall to compete with the other huge players (such as Electricité de France, which has begun acquiring shares of companies in other countries in northern Europe). This second argument would be persuasive if the facts were other than they are now. Currently, Vattenfall is still the dominant firm in the Swedish market, and the prospect of large-scale foreign penetration into the Swedish electricity market seems presently quite remote.

Hence, it would be ironic if the Swedish deregulation plan adopted not only the admirable innovations but also the avoidable flaws of the British deregulation experiment. If Sweden does continue with its approach to deregulation of the electricity market, consideration should be given to whether Vattenfall should be divided into two or three production companies. In addition, the continuation of state ownership of Vattenfall in the deregulated environment seems pointless. State ownership is likely to produce muddled objectives, political intervention, and subsidized investments. In the past, Vattenfall has clearly benefited from the low cost of capital available to government agencies, leading to a misallocation of capital between private and public firms and to an overcapitalized electricity generating system. Privatization would cure both of these ailments.

Additionally, one of the major differences between the Swedish case and the British and American deregulation cases is that the Swedish electricity industry appears not to be saddled with a significant quantity of "stranded assets." This term refers to the potential for high-cost plants to remain open under rate-of-return regulation because the monopoly positions of their owners allow them to pass the high costs through to customers. In a competitive market, any plant whose marginal production costs were higher than the price would be shut down; in a regulated monopoly, a high-cost plant may remain open if the sunk and stranded costs can thereby be recovered. In Britain and the United States, many of the nuclear power stations are stranded assets, and in Britain and Germany the coal mines have until recently been another set of stranded assets that regulators and fiscal authorities have attempted to keep open. Efficiency would dictate closing these stranded assets, but the political difficulties have until recently proven insurmountable.

Data are incomplete on the question of whether the nuclear power plants in Sweden are stranded assets (that is, whether they would be shut down in a competitive market). Estimates that we provide below indicate that the nuclear power plants currently in operation will continue to be economical in the future, particularly if prices rise to the level of the cost of replacement power. Both nuclear power plants and hydroelectric power plants currently have very low marginal costs of production—on the order of 0.7 to 2.0 cents per kilowatt-hour (kWh) compared to around 7 cents per kWh for replacement power (all in 1995 prices) according to estimates presented below. It seems likely, therefore, that the issue of stranded costs of nuclear power will *not* be a major obstacle to efficient electricity deregulation in Sweden.

In summary, the Swedish deregulation proposal appears to be well structured to produce a competitive market in generation. The major reservation about the proposal is retaining Vattenfall as a dominant market force. Consideration should be given to whether Vattenfall should be privatized and broken into two or three different entities to make the electricity market effectively competitive.

Deregulation and the Nuclear Phaseout. What does the advent of a more competitive electricity market imply for the impact of a phaseout of nuclear power in Sweden? The major surprise is that there is likely to be little relationship between the two. The reason is that the cost of a nuclear phaseout is determined primarily by the replacement cost of the power; furthermore, that cost is likely to be largely independent of the regulatory regime. It follows that deregulation is likely to have little impact upon the economics of nuclear power in Sweden.

To see this, suppose that 10 tWh of nuclear generation is removed from the Swedish electricity system. The economic cost of this measure would be the cost of the replacement power or of the conservation measures that would be induced by the higher prices. Unless the prices or supply elasticity were quite different under the different regulatory regimes, the cost of the 10-tWh loss in production would to a first approximation be the same in a regime with club-type regulation as in a competitive market.

There are, however, two second-order consequences of deregulation, and, on the whole, these are likely to make the cost of the nuclear phaseout lower in a competitive market than in a regulated or semimonopolistic industrial structure. For purposes of this discussion, I will assume that the deregulated market behaves in a competitive manner. In this case, the effects of a nuclear phaseout will be different in alternative regulatory regimes because of differences in the market price of electricity and because of differences in robustness to shocks. Again, recall that we are analyzing the *difference in the impact of a nuclear phaseout* between the regulated and deregulated industrial structures, not the cost of the phaseout itself.

The first difference arises because of the impact of deregulation on the market price of electricity and the marginal cost of nuclear power. In fact, the impact of deregulation is ambiguous. The general presumption is that the electricity price will be lower in the deregulated market than in the regulated market because inefficient production and consumption will be discouraged. If deregulation makes nuclear power more efficient and lowers its marginal costs without changing the electricity price, then a nuclear phaseout would actually be *costlier* in a deregulated environment. On the other hand, if deregulation reorganized the industry and lowered the electricity price without changing the marginal cost of nuclear power, then a nuclear phaseout would be less costly in a deregulated environment.

A hypothetical example will clarify this paradoxical result. Suppose (contrary to the facts in Sweden) that the nuclear power plants were uneconomical "stranded assets." In a deregulated environment, they would therefore be shut down and decommissioned, and the firms might well experience financial difficulties. In this case, a nuclear phaseout would have no costs because the plants would be shut down in any case. Contrast this with a regulated environment where firms can pass through the high costs to their customers and would keep the plants open to recover their capital costs. In this case, a nuclear phaseout would actually *save* money because it would be retiring uneconomical plants. This first difference between the regulated and the deregulated environments is likely to be small but is in principle ambiguous.

A second difference arises because of differences in robustness to shocks. The general presumption, bolstered by much historical evidence, is that a competitive environment is likely to be more robust to shocks (or technically have more elastic supply) than a regulated system. This is so because shocks can be spread more widely across different suppliers and users. If Sweden has a dry year or if demand is high because of an economic recovery, then prices will rise less sharply in Sweden if it is part of a broader Scandinavian or European electricity market than if it were isolated. In the limit, if Sweden becomes a classical small country in the electricity market, electricity prices will be determined in the broader electricity market and will become independent of demand-and-supply conditions in Sweden. In general, we would expect that the higher robustness of a deregulated environment would reduce the cost of a nuclear phaseout. In a more robust environment, a phaseout would cause a smaller electricity price rise and consequently less real income loss.

The summary view on the Swedish deregulation plan is that it seems by and large headed in the right direction—with the notable exception of allowing a dominant firm to exercise market power in the electricity market. However, there seems to be very little relationship between the regulatory regime itself and the merits or demerits of a nuclear phaseout. We will present below estimates of the impacts of the regulatory regime on nuclear phaseout costs, but we would on general principles expect these to be small relative to the direct costs.

ENDNOTES

[1]This review is based on Organisation for Economic Co-operation and Development, Nuclear Energy Agency, *Nuclear Power and Public Opinion,* Paris, OECD, 1984.

[2]A thorough discussion of most aspects of climate change is contained in U.S. National Academy of Sciences, Committee on Science, Engineering, and Public Policy, *Policy Implications of Greenhouse Warming: Mitigation, Adaptation, and the Science Base,* Washington, D.C., National Academy Press, 1992. A thorough survey of the science, full of interesting figures and background, is contained in Intergovernmental Panel on Climate Change, J.T. Houghton, G.J. Jenkins, and J.J. Ephraums, eds., *Climate Change: The IPCC Scientific Assessment,* New York, Cambridge University Press, 1990. The most recent report by the IPCC indicates that the pace of warming will be less rapid than was estimated in the earlier report but finds stronger evidence of a greenhouse warming "signal" (see Intergovernmental Panel on Climate Change, Working Group I, J.J. Houghton, L.G. Meiro Filho, B.A. Callander, N. Harris, A. Kattenberg, and K. Maskell, eds., *Climate Change 1995: The Science of Climate Change,* New York, Cambridge University Press, 1996.

[3]Integrated assessment models are attempts to combine all the major components of greenhouse warming; an early example is discussed in William D. Nordhaus, *Managing the Global Commons: The Economics of Greenhouse Warming,* MIT Press, 1994. A survey of models is currently under way under the aegis of the Stanford Energy Modeling Forum 14 (EMF-14) led by Stanford's John Weyant. The statement in the text about integrated assessment models refers to the ten or so models surveyed in an early round of the EMF-14.

[4]The most careful studies of the impact of greenhouse warming have been conducted for the United States, and a thoughtful review is contained in U.S. National Academy of Sciences, Committee on Science, Engineering, and Public Policy, *Policy Implications of Greenhouse Warming: Mitigation, Adaptation, and the Science Base,* Washington, D.C., National Academy Press, 1992. Several international assessments of the economic consequences of climate change are currently under way; see particularly Intergovernmental Panel on Climate Change, Working Group III, James P. Bruce, Hoesung Lee, and Erik F. Haites, eds., *Climate Change 1995: Economic and Social Dimensions,* New York, Cambridge University Press, 1996.

[5]See Greenland Ice-Core Project (GRIP) Members, "Climate Instability during the Last Interglacial Period Recorded in the GRIP Ice Core," *Nature,* July 15, 1993, pp. 203–208.

[6]See M.G. Morgan and D. Keith, "Subjective Judgments by Climate Experts," *Environmental Science and Technology*, Vol. 29, No. 10, pp. 468A-476A, 1995.

[7]Government of Sweden, Ministry of the Environment and Natural Resources, *Sweden's National Report Under the United Nations Framework Convention on Climate Change,* September 1994 (hereafter "Sweden's National Report").

[8]Accounting for environmental emissions is likely to be a major growth industry in the coming years. I have denoted emissions as "counted" those that are included in the national targets, these being industrial emissions of carbon dioxide. What is in and what is out are quite controversial, however. The major loophole is the exclusion from counted emissions of those embodied in foreign trade.

[9]See NUTEK (Närings-och Teknikutvecklingsverket), *Scenarier över Energisystemts utveckling år 2020 (del II)*, PM 95-03-02, Tabell 5.3.

[10]See Lennart Hjalmarsson, "From Club Regulation to Market Competition," in Richard Gilbert and Edward Kahn (eds.), *International Comparisons of Electricity Regulation,* New York, Cambridge University Press, 1996.

[11]This discussion is based primarily upon John Vickers and George Yarrow, "The British Electricity Experiment," *Economic Policy,* April 1991, pp. 188–232; and David Newbery and Richard Green, "Regulation, Public Ownership and Privatisation of the English Electricity Industry," in Richard Gilbert and Edward Kahn 1996, ibid.

[12]For a discussion of the major liberalization plans, see Richard Gilbert and Edward Kahn 1996, ibid.

[13]This summary is based on *Summary of Government Bill 1993.94:162: A Competitive Electricity Trade,* Stockholm, 1994, and from discussions with government and industry experts.

[14]This point is effectively argued in Lennart Hjalmarsson, "From Club Regulation to Market Competition," in Richard Gilbert and Edward Kahn 1996, ibid.

[15]Lars Bergman and Bo Andersson, "Market Structure and the Price of Electricity: An Ex Ante Analysis of the Deregulated Swedish Electricity Market," Stockholm School of Economics, mimeo, September 1994, p. 13.

PART II
Options and Impacts

4
Building Blocks

This chapter's first section discusses alternative projections of output and the growth in the electricity market. The next section provides a summary analysis of how a nuclear phaseout will affect the Swedish electricity market.

SUPPLY AND DEMAND SITUATIONS IN 1990 AND 2010

The economic and environmental impact of a nuclear phaseout will depend on the trajectory of demand and alternative sources of electricity supply over the next two or three decades. If electricity demand rises sharply, as it has over the last quarter century, and in the absence of revolutionary breakthroughs in alternative and environmentally benign sources of electricity, then a nuclear phaseout will require significant investments in alternative sources of supply (either in Sweden or in an exporting country) or a major increase in electricity prices to curtail demand. To the extent that new sources of supply or energy conservation are less costly than currently anticipated, the nuclear phaseout would then have lower cost for Sweden.

Background

One of the fundamental philosophical assumptions that motivate this study is that it should analyze the Swedish dilemma using state-of-the-art methodologies and rely to the maximum possible extent on assumptions independent of those that are the "conventional wisdom" in Sweden. It is hardly helpful for a foreign observer simply to absorb Swedish assumptions and prejudices—that could much better be undertaken by someone more familiar with the country and its psychology. To ensure to the maximum possible extent an independent vantage point, I have constructed the estimates and models used here and laid out the economic and electricity assumptions from first principles rather than relying on "standard assumptions." This approach, of course, can easily run aground if distance breeds ignorance, but at the least it will provide an alternative perspective.

One of the main uncertainties in the projections is the growth of electricity demand. It is important to understand the technical definition here: *Electricity demand* is the demand for electricity that would occur at constant relative prices of electricity and other energy sources. This is different from the *quantity of electricity demanded*, which will be larger or smaller than electricity demand, for example, if the price of electricity declines or increases over time, respectively.

In this study, I have derived two independent estimates of the growth of electricity demand. The first one, described in this section, is a pure demand model in which I make certain assumptions about the state of the electricity market and then estimate the growth in the quantity of electricity demanded. It is incomplete and is used mainly for illustrative purposes. The second set of projections is developed in the section on the Swedish Energy and Environmental Policy (SEEP) model. That second approach is a "general equilibrium" approach in which supply and demand for electricity and nonelectric energy are integrated with the rest of the economy. The second approach is more complete but also more complex. In addition, for comparative purposes, I will contrast the energy use assumptions in the two approaches developed here with those developed recently in the Swedish context.

The Econometric Demand Approach

In the first approach, which relies largely on econometric techniques, I have estimated demand functions for both the residential and industrial electricity sectors of the Swedish economy. The methodology in brief is the following:[1]

The first question involves the growth of Swedish real gross domestic product (GDP) over the next quarter century. To derive these estimates, I applied standard techniques to historical data. I began by obtaining population projections and used them to estimate the Swedish labor force from 1993 to 2010. To obtain employment, I assumed that the unemployment rate would decline gradually, reaching 4% in 2010.

I next estimated productivity functions for the period 1970–1993. Productivity was defined as real GDP per employed worker. The estimated equation used the logarithm of real GDP per worker as a dependent variable with time and the unemployment rate as the independent variables, and added a first-order autoregression. Using this relationship, I then projected real GDP on the basis of projected productivity and employment.

The results of the GDP regression are shown in Part A of Table 1 in the row labeled *Econometric*. The estimated growth in Swedish real GDP over the 1993–2020 period is estimated to average 2.24% per year. That rate is considerably higher than has been experienced in the last two decades. This projection is based on an estimate that productivity growth will be 1.3% per year over the period until 2020, which rate is intermediate between the dismal experience of the 1970s and the more satisfactory rate of the 1980–1993 period. The major reason for the high GDP growth in this production, however, lies in its assumption that Sweden will eventually recover from the deep depression that currently prevails. Looking at subperiods, the growth rate is projected to average 2.5% annually over the 1993–2005 period and slow to the estimated potential growth rate of 1.75% annually after 2010.

Estimating electricity demand is always a tricky exercise. For the purpose of the econometric study, we need estimates of the income and price elasticities of the demand for electricity along with assumptions about income growth as well as price trends. There are numerous studies of energy and electricity demand for other countries as well as for Sweden, and it was thought useful to develop independent estimates for this project. In estimating the demand for electricity, we separated small users (residential and commercial) from large users (industrial). We then estimated demand over the 1970–1993 period for each sector as a function of GDP, the price for that segment of demand, and prices of alternative fuels or the

Table 1. Alternative Projections for GDP, Electricity Prices, and Electricity Demand.

Part A. Output Trends (real GDP growth, average annual growth rate)

Source	*1970–80*	*1980–93*	*1993–2020*
Actual	1.96	1.14	na
Projected			
NUTEK	na	na	1.70
SEEP baseline	na	na	1.95
Econometric	na	na	2.24
Ministry of Finance[a]	na	na	1.95

Part B. Electricity Use, Supply, and Price

		Electricity price (cents per kWh, 1995 prices)		
Year	*Source*	*Residential*	*Industrial*	*Electricity demand (terawatt-hours)*
1970	Actual	6.7	4.2	63.4
1993	Actual	9.0	4.2	140.4
2010	NUTEK	na	na	na
	SEEP baseline	12.4	6.8	142
	Econometric[b]	10.6–18.4	5–10	155–196
2020	NUTEK	15.3–18.6	5.9–9.3	134–158
	SEEP baseline	12.9	7.3	165
	Econometric[b]	10.6–18.4	5–10	176–223

Note: The projections are from Swedish sources for actual data, for NUTEK projections, and for the Ministry of Finance. Estimates for the SEEP baseline are presented later in this study, while those of the "econometric" estimates are discussed in the text. na = not applicable or not available.

[a]Through 2010 only.

[b]Estimates show the range of electricity use, with Case III being the high price and low demand figure while Case I is the low price and high demand number.

general price level. In addition, a lag structure was introduced to allow for the slow adjustment of electricity use to changing prices. In addition, instrumental variable estimators were used to test for the potential influence of simultaneous-equation bias.

The results of the energy demand study were as follows: For the industrial sector, demand was determined to be quite inelastic, with a short-run price elasticity with respect to industrial electricity price of -0.15 and a long-run price elasticity of -0.17. Industrial demand was found to be highly sensitive to output growth with a long-run income elasticity of 1.14. For the residential sector, demand was determined to be more price sensitive with a price elasticity of -0.13 through the end of the first year and a long-run price elasticity of -0.54. These estimates are consistent with most other studies in showing relatively price-inelastic demand for electricity for periods of up to a decade.

Using the GDP projections and the energy demand estimates, we then projected electricity demand to the year 2020 under three price scenarios. In all scenarios, GDP growth followed the projection discussed above. In addition, nonelectric energy prices were held constant (relative to the Swedish retail price index) at their 1993 levels. The three projections (designated I, II, and III) differed according to the assumption about the trend price of electricity. They were (I) that electricity prices were held constant at 1993 levels, (II) that electricity prices increased smoothly by 50% between 1993 and 2010 and then were held constant at that higher level after 2010, and (III) that electricity prices increased smoothly by 100% between 1993 and 2010 and then were held constant at that higher level after 2010. Case I would correspond to an optimistic case where demand was very low or breakthroughs in supply allowed new generation at quite low costs; Case II is probably close to a reasonable projection of electricity prices over the next two decades; Case III corresponds to what might happen with relatively rapid demand growth and severe supply constraints on electricity production. It is unlikely that the growth rates would be at equal rates in the different sectors, but this assumption is at this stage primarily for illustrative purposes.

Part B of Table 1 shows the range of estimated prices and growth in electricity use for the three "econometric" scenarios along with the other estimates to be reviewed shortly. According to these projections, we should expect relatively strong growth in electricity demand in Sweden over the next quarter century. In none of the cases would we expect to see a decline in electricity demand, and in the high-growth case electricity demand is expected to grow by 60%. In the low-growth case, where electricity prices double in real terms over the next two decades, electricity demand is expected to grow by about 25% over the period from 1994 to 2020.

Of course, like most demand projections, these are conjectural and full of the potential for surprise, but they do suggest the need for substantial increases in supply (whether from domestic generation or from electricity imports). This allows us to estimate the need for sources of supply with and without the nuclear phaseout. Even with continued deployment of the existing nuclear capacity, the low "econometric" case suggests that capacity must be expanded by between 15 and 55 tWh (terawatt-hours or billion kilowatt-hours [kWh]) (or between 10% and 40% of 1993 generation) between now and 2010. If nuclear power is discontinued, this would imply an increase of an additional 70 tWh for a total increase of 85 to 125 tWh, representing more than a doubling of current nonnuclear capacity between now and 2010. Unless these projections are far off base or, alternatively, unless electricity prices rise even more sharply than is anticipated in Case III, Sweden will need to find a substantial new supply over the next fifteen years if it chooses to phase out its nuclear power.

Alternative Projections

How do these projections compare with others? A 1995 study by a Swedish energy agency (NUTEK) made a number of projections for use by a group of experts that was asked by the Swedish government to study Swedish energy policy.[2] The NUTEK study projected a slower growth in GDP at a rate of 1.7% annually rather than the 2.2% derived in our energy demand estimates. A more thorough recent government study of the long-term prospects for the Swedish economy, by contrast, projects a growth rate of 1.95% per year over the period 1994–2010.[3]

The next section presents the SEEP model, which is a mathematical programming model of the Swedish energy system. This model calculates energy and electricity demand from the fundamentals of technology and preferences and is calibrated to 1994 levels of prices and outputs. The results of the SEEP model are shown in detail below (see particularly Table 1 of Chapter 5 and Tables 1, 2, and 3 of Chapter 7), but Table 1 of Chapter 5 (see page 78) displays some of the major results for comparative purposes. We have used intermediate estimates of the growth of GDP in constructing the SEEP model—1.95% annually rather than the lower number in the

NUTEK study or the higher number we estimated in our econometric study described in the last section.

With regard to electricity prices, these are projected to increase sharply in the NUTEK scenarios. For NUTEK's Scenarios 1, 2, and 3, the electricity price for industry is projected to increase between 78% and 183% (in constant prices), while for residential users the price is projected to increase between 38% and 92% (again, in constant prices). These compare with the range of between 0% and 100% for the price increase in the econometric electricity-demand projections discussed above.

Prices in the SEEP model are estimated to rise relatively modestly, with an increase of approximately 30% in the wholesale (bus bar) price of electricity over the 1994–2020 period in the case where nuclear power is not phased out. Electricity demand in the SEEP model is estimated to grow more slowly than in the econometric estimates presented here, with the estimated consumption in the baseline case being 142 tWh in 2010 and 165 tWh in 2020 as opposed to the range of 155 to 196 tWh for 2010 and 176 to 223 tWh in the econometric estimates provided here. A final comparison is the estimates in the SEEP model presented below with those in the NUTEK study. The SEEP model projects electricity demand of 142 tWh in 2010, which is comfortably in the range of estimates that are provided by NUTEK.

In conclusion, the econometric study that was undertaken for this study indicates higher output growth and higher growth in electricity use than is commonly assumed in Swedish energy studies. For the central purposes of this study, it turns out that these assumptions are not crucial to the major results. It should be noted, however, that we have developed the estimates of demand and output growth on assumptions that appear to use lower growth rates than are consistent with the econometric estimates developed here.

IMPACTS OF SUPPLY REDUCTIONS

Later in this study, I present a number of estimates of the cost of a nuclear phaseout. At this stage, it will be useful to show the economics of the nuclear phaseout diagrammatically. This is best accomplished using an elementary supply-and-demand approach with illustrative numbers for the electricity market.

Figure 1 shows the supply and demand for electricity in Sweden with nuclear power. The demand curve (*D*) has the usual downward slope to reflect the dependence of quantity demanded, *Q*, on the price of electricity, measured on the vertical axis and denoted P_{elect}. The supply curve is drawn on the assumption that the marginal cost of existing capacity is MC_{exist}. Total existing capacity as of today is Q_{all}. When existing capacity is reached (at the kink

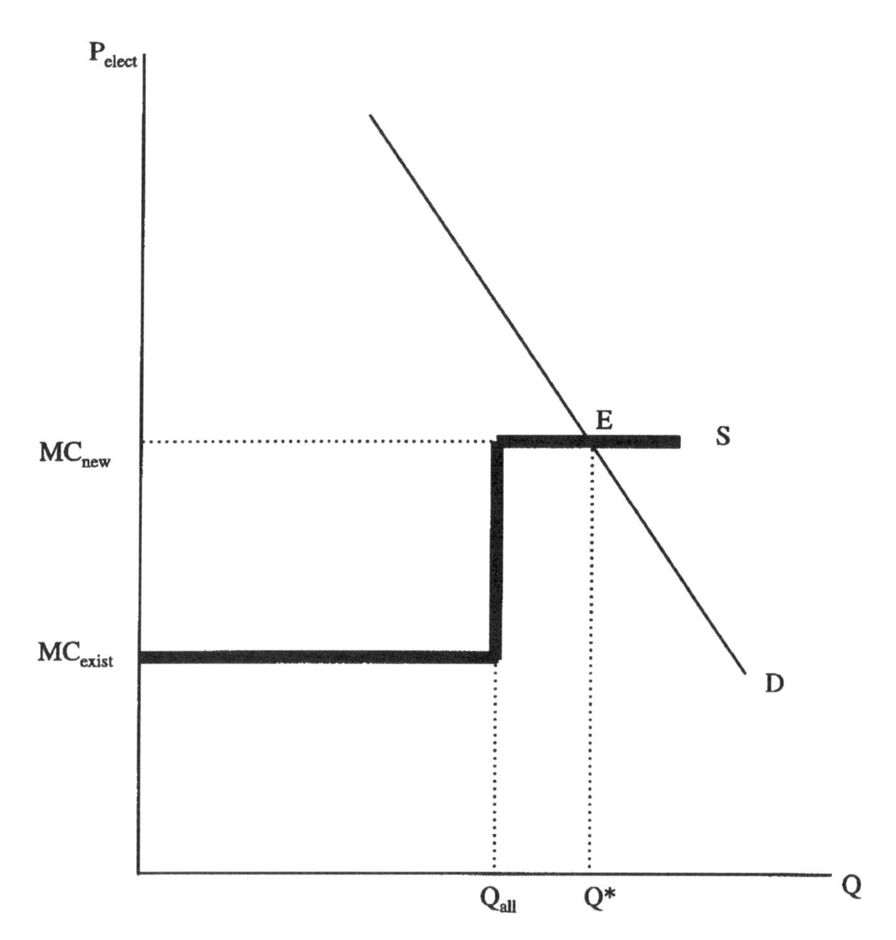

Figure 1. Swedish Supply and Demand for Electricity with Nuclear Power.

in the supply curve at Q_{all}), any further demand must be met with new capacity, which has marginal cost equal to MC_{new}. Hence, the supply curve with nuclear power is the heavy kinked line in Figure 1.

Figure 1 might show the electricity market in the year of the nuclear phaseout, say in 2010. At that time, existing capacity is insufficient to meet the demand, so a small increment in new capacity (equal to Q^* minus Q_{all}) must be added at the marginal cost of MC_{new}. The market price of electricity in a competitive market will be equal to the marginal cost of new capacity, given by MC_{new}, and the quantity demanded is given by Q*.

Figure 2 then illustrates the impact of a nuclear shutdown. To show the effects, we assume that current nonnuclear capacity is Q_{all} minus Q_{nuc}, while Q_{nuc} is current nuclear capacity. A nuclear shutdown will affect the industry by removing production of Q_{nuc}, which is low-marginal-cost capacity. This will have the effect of shifting the supply curve to the left by the nuclear capacity, Q_{nuc}. Hence, in Figure 2, the electricity supply curve with the nuclear capacity is *ABCGE*, while the modified supply curve after the nuclear phaseout will be *ABFGE*.

In the pictured case, it is extremely simple to calculate the cost of a nuclear phaseout; this case assumes that the quantity demanded is always greater than the existing capacity. In this case, price will always be equal to the marginal cost of new capacity, and total generation will be independent of demand. While this is oversimplified, it does provide an accurate estimate of the phaseout cost.

In this simplified case, the cost of the phaseout is simply the cost increase due to the phaseout. This is easily depicted in Figure 2 by the shaded area *BFGC*. A little reflection will indicate that *BFGC* equals the marginal cost differential (MC_{new} minus MC_{exist}) times nuclear capacity, X_{nuc}. This sum is equal to the difference in marginal cost between new capacity and nuclear power times the nuclear power generation. *In other words, the annual cost of a nuclear shutdown is the annual nuclear generation times the cost differential between the average cost of the new generation and the marginal cost of existing nuclear power*. In other words, the cost of the nuclear shutdown is simply the cost of replacement power.

We can easily make an order-of-magnitude estimate of the cost using the simplified approach in Figure 2. Estimates prepared later

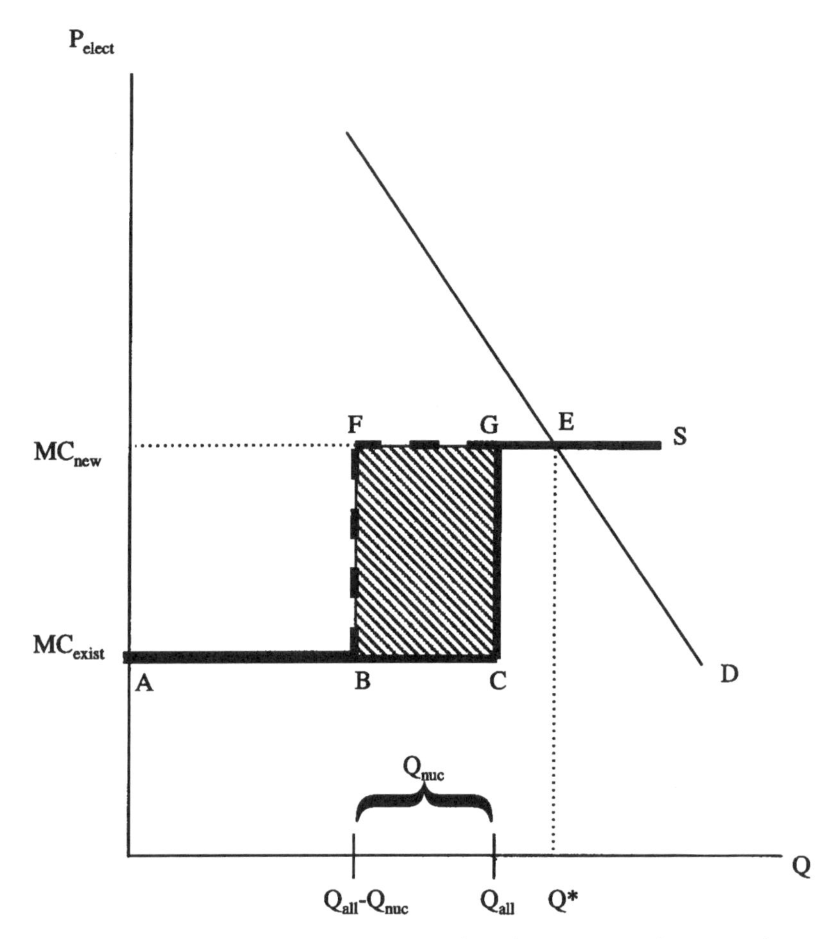

Figure 2. Swedish Supply and Demand for Electricity without Nuclear Power.

in this study indicate that the marginal or avoidable cost of existing nuclear power is around 1.7 cents per kWh, while the marginal cost of new generation is around 5 cents per kWh (all in 1995 prices converted at an exchange rate of 7 kronor per U.S. dollar). Nuclear generation for the last four years has averaged around 70 tWh. Therefore the annual cost of a shutdown would be $(0.05 - 0.017) x 70 billion per year = $2.3 billion per year.

To obtain a present value, we need to know the useful lifetime of the nuclear plants as well as the appropriate discount rate.

Table 2. Estimated Costs of Nuclear Shutdown in Simple Supply and Demand Model (present value of nuclear shutdown in 2010 in billions of U.S. dollars, 1995 prices).

Lifetime (years)	*Discounted to shutdown date of 2010*	*Discounted to 1995*
10	18.7	9.0
20	30.2	14.5
30	37.6	18.1

Note: This table shows the estimated costs of a nuclear shutdown occurring in the year 2010 in the simple supply and demand model. The lifetimes represent the expected number of years of operation as of the date of the shutdown. For this example, we assume that there is no price response and that the difference between the marginal cost of the replacement power and the avoidable cost of existing nuclear power is 3.3 cents per kWh. The estimates are in 1995 prices. The second column is the present value discounted back to the year of the shutdown. The third column takes this and discounts the value back to 1995; it thus divides the second column by $(1.05)^{15} = 2.08$, which is the discount factor 1.05 compounded from 1995 to 2010.

Assume that the discount rate is 5% per year. This is a real discount rate, correcting for the movement in the price level. Then, for economic lifetimes ranging from ten to thirty years following the shutdown, Table 2 shows the estimated present value of the cost of a nuclear shutdown. The first column shows the present value where the present value is discounted to the date of the shutdown; the second column shows the present value where the shutdown takes place in 2010 and the value is discounted back to 1995.

The estimate of the cost of a nuclear shutdown that reduces the lifetime by ten years in the simplest supply-and-demand model is seen to be $9 billion (in 1995 prices discounted to 1995). This number is somewhat misleading, however, because the costs *at the time of the shutdown* are considerably higher, $19 billion (again in 1995 prices). Lifetimes of twenty and thirty years have somewhat less than proportional increases in the present value of the cost, again because of the discounting. It is useful to compare this simplified example with the full model presented below. In the SEEP model, we estimate that the economic cost of a nuclear shutdown in 2010 is $8.9 billion as opposed to the $9.0 billion in the simple model.

We will find subtle differences between this example and the approaches that will be discussed below, but this calculation shows that the basic logic introduced in Figures 1 and 2 captures the full economic impact remarkably well. The basic logic of this example is straightforward. As long as consumption at the shutdown date is greater than total capacity, then (under the simplified assumptions about the cost structure) the electricity price—and therefore the consumption—will be unaffected by the shutdown. In this case, we simply need to calculate future nuclear production and the cost differential cost between replacement power and nuclear power to determine the cost of the phaseout.

There are a number of sensitivity analyses one can perform on this elementary calculation taking alternative values, say, of the discount rate, of nuclear generation capacity, of marginal costs, and of demand functions. These will be forgone at this point given that we will be presenting more elaborate models discussed below. The basic results presented here will carry through the more elaborate discussion to follow.

ENDNOTES

[1]The estimates of Swedish electricity demand growth were undertaken with the research assistance of David Lam of Yale College.

[2]See NUTEK (Närings-och Teknikutvecklingsverket), *Scenarier över Energisystemts utveckling år 2020 (del II)*, PM 95-03-02.

[3]Government of Sweden, Ministry of Finance, *The Medium Term Survey of the Swedish Economy*, Stockholm, 1995, p. 223

5

The Swedish Energy and Environmental Policy (SEEP) Model

Once we go beyond the simplest supply-and-demand examples, the interaction of energy, environmental, and economic issues becomes extremely complex. It is impossible to hold in the normal human mind all the different variables that are necessary to calculate the impacts of different policies under alternative scenarios and with a variety of possible growth paths. Therefore, in order to understand the implications of alternative energy and environmental policy issues in a more realistic fashion, I have constructed an economic model of the Swedish energy-environment system that will provide estimates of the economic impacts of alternative policies and approaches. I call the model the SEEP model, or Swedish Energy and Environmental Policy model. I begin with a nontechnical description of the SEEP model. The next section provides a more detailed description, while later sections discuss the calibration of the model.

The SEEP model relies on modern economic theory to integrate the economic, energy, and environmental concerns into a unified treatment. The model represents the entire Swedish economy at different levels of detail. The nonenergy sectors are represented as an aggregate. The energy sector is broken into three different segments: specific electric, transportation, and other energy uses (heat, boilers, and so forth). In addition, an environmental sector is added that can incorporate constraints on emissions such as carbon dioxide and sulfur dioxide.

The energy market has coupled the supply and demand sides. Demand responds to total output, the prices of different fuels, and the prices of different end uses. The demand for energy is not derived from an econometric model but is a derived demand based on a production function. Supply depends on fuel and capital-goods prices, the real interest or discount rate, government environmental and nuclear regulations, and existing capacities. The two sides of the market are balanced by prices; that is, prices are determined so that desired purchases and desired sales are equal. At the market equilibrium, all production processes must break even or shut down, and prices are determined on the basis of marginal costs, not average costs.

The most important part of the model concerns electricity production. For this, we represent production by distinct technologies (nuclear, hydroelectric, coal-fired, and so forth) using a linear-programming framework. Current capacity has relatively low operating costs with a limited capacity, while new capacity is unconstrained but will have higher production costs. The model is simplified by omitting such factors as constraints on the expansion of new technologies, but this simplification will have little effect on the estimated cost of a nuclear shutdown.

The model is a dynamic one that runs in annual steps from 1994 out through a time horizon that is the maximum lifetime of current nuclear plants, taken to be 2040. The choices about consumption, energy demand, fuel use, and other economic variables are determined by assuming that Sweden maximizes its economic welfare. In technical language, economic welfare is the sum of nonenergy net output plus consumers' surplus in the energy sector less the costs of energy production. Consumers' surplus is estimated using standard economic demand analysis. The costs of production are both imported and domestically produced energy (with a special focus on electricity).

The SEEP model is a "calibrated" model rather than an econometric one. It is calibrated by setting the parameters in such a way that the outcomes match the actual data for Sweden for 1994, or in some cases 1993. Some of the calibrations are straightforward, such as those involving gross domestic product (GDP), energy prices, energy production, and fuel use. Other calibrations are more complicated, particularly those involving the production relationships.

In addition, because the model requires future values of the variables to make projections, it is necessary to estimate future values of energy and economic variables. For purposes of estimating the impacts of a nuclear phaseout, the most important assumptions involve the cost of replacement power, the capacity and lifetime of nuclear plants, and the avoidable costs of current nuclear plants. Other variables, important in other contexts but less crucial here, involve the growth of GDP, the price and income elasticities of demand for energy, and the industrial organization and foreign trade in electricity.

Once the model has been calibrated, we can then examine alternative assumptions to test the impact of alternative policies or external conditions upon Swedish economic welfare. The major policies examined here are different options on nuclear power, such as early phaseout, late phaseout, and no phaseout. In addition, models of this kind are extremely useful in allowing a determination of the combined impacts of nuclear power policy with other policies, such as climate-change policy, protecting rivers, or the regulatory regime. Of course, no model can provide a perfect forecast of future events. Its strength, rather, is to provide a way of testing the implications of different assumptions about economic events and policies on energy, economic, and environmental outcomes.

The next section describes the structure of the SEEP model in greater detail. The subsequent sections then describe the derivation of the critical parameters, such as costs of nuclear power and other electricity sources, lifetimes, and environmental costs.

STRUCTURE OF THE MODEL

In this section, I describe the structure of the SEEP model. (The actual computer program is available from the author on request.[1]) The model is a standard optimization model that maximizes the discounted value of Swedish "utility" (which is a technical term for the consumer value of produced goods and services). The model operates over the time horizon of 1994–2040. Output is divided into a nonenergy sector and an energy sector. (The model has been calibrated and was run on the GAMS software package[2] using an Intel

Pentium 66 processor. A standard run takes about four minutes to solve.)

The energy sector is divided into three components: one that is devoted to specific electric purposes, one that is devoted to transportation, and one that is for other energy uses. The three components of energy demand are combined in a log-linear (or "Cobb-Douglas") utility function in which the shares are equal to expenditures in the base year, 1994. The model is calibrated so that the level of utility is equal to Swedish GDP in 1994, and the real discount rate is set at 5% per year. For the baseline run, we have adopted government medium-run projections in which real GDP is estimated to grow at 1.95% per year from 1994 until the end of the estimation period.[3]

Production is differentiated for the three energy sectors. The treatment of electricity use is critical and will therefore be described in greater detail than the rest of the model. Electricity use is divided into two parts.[4]

- One source of demand is called the *specific electricity uses*. These uses—which include electricity going to electric motors, lighting, electrolysis, and similar categories—are ones in which electricity is the only realistic technology for meeting the demand efficiently.
- In addition, there is a second sector, *general electricity uses,* such as space or general heating and other uses, for which a wide variety of different technologies can be adopted.

A substantial part of electricity use in Sweden falls into this second category—notably electric heating and industrial boilers. For this second demand segment of electricity demand, it is assumed that these end uses can be produced with electricity, oil, coal, gas, and biofuels. In these sectors, the costs of production will vary for the different fuels, but most of the demand could be met by alternative fuels without an enormous penalty in cost or efficiency.[5]

This partition of electricity demand into the two segments is critical to understanding trends in Swedish electricity demand. In the 1980s, as the electricity industry experienced large increments in supply with the major increase in nuclear power capacity, demand increased in large part because of an expansion in the general electricity uses in electricity heating and industrial boilers. As electricity prices rise in the near term, electricity use in these sectors will be reduced, making way for a relatively larger fraction of the electricity supply to go to the specific electricity segment. Hence, electricity demand will have one segment—the general electricity uses—that has a relatively high price elasticity of demand (with other fuel prices held constant), while the other segment has a relatively low price elasticity of demand because there is no economical alternative energy source.

The specification in the two other energy sectors is straightforward. For the transportation sector, only petroleum products can meet the demand. For the other energy uses, energy needs can be met with electricity, biofuels, coal, gas, or petroleum. In the latter sector, the different fuels are not perfect substitutes; rather, there are ample substitution possibilities as represented by partial elasticities of substitution of two between the different energy sources.

The demand for electricity and other energy products is modeled in an optimization framework rather than a behavioral framework. That is, demand is the outcome of profit maximization subject to the production possibilities and prices in the marketplace, rather than as an econometric demand function. Hence, the demand for electricity will arise from the specific electric and general electricity uses and is a function of total output, the prices of the different fuels, and the production functions that combine inputs into outputs and different fuels into heat or light.

The demand functions in the SEEP model are therefore different from the econometric demand estimates presented in the earlier section and rely instead on estimates of the production functions. This implies that the price elasticities are relatively high for the general electricity uses, such as electric heating or boiler uses, and relatively low for specific electricity uses, such as lighting. Under this specification, the demand elasticities for the three components (with respect to final prices inclusive of distribution costs and taxes) are approximately -1.0, while the income elasticities are all unity.[6] These elasticities are somewhat higher than conventional estimates and are considerably higher than those derived in the section on demand above. On the other hand, econometric estimates have had great difficulty modeling the full substitution possibilities in a way that is compatible with the underlying structure of energy technology.

Production of electricity begins with existing sources (primarily nuclear and hydroelectric), which have relatively low marginal costs. Additional electricity production is in principle allowed from new gas-fired, coal-fired, and nuclear generation. No additional hydroelectricity is included in the base case, nor is there any provision for incremental additions to imports in the base case. In addition to the costs of production of the different sources, retail electricity prices include distribution costs and fuel taxes (at 1994 levels). Fuel taxes are assumed to reflect true social costs, either because they reflect social valuation of externalities associated with the good or service or because they reflect the marginal deadweight loss associated with raising revenues. We have in the base run omitted environmental taxes. These have been omitted because they have been so frequently changed in recent years and because in some cases (carbon dioxide, in particular) they do not appear to be closely related to the relevant externalities. All prices and costs are set equal to 1994 levels and are measured in 1995 prices. New generating costs are estimated from materials derived from international agencies and are described in detail below.

The environmental sector is represented by three constraints, with respect to hydroelectric power, carbon dioxide (CO_2) emissions, and sulfur dioxide (SO_2) emissions. For hydroelectric power, it is assumed that the production of hydroelectric power is limited to current levels (although no random elements from precipitation are included in the analysis). We include explicit accounting for energy-sector emissions of SO_2 and CO_2, and for a number of runs we constrain emissions to consider the impact of implementing international agreements on climate change. We do not include the external costs of the environmental externalities in the base case. These are omitted because estimates of the costs are so uncertain at this stage.

The SEEP model clearly has shortcomings as a representation of the highly complex energy system. It lacks the dynamics of investment and capital stocks; it omits the details of region, time-of-day, and time-of-year demands; the weather is not considered, nor are the interactions of Sweden with other countries. In addition, the economy is represented as a smoothly functioning system without business cycles, unemployment, exchange-rate crises, tax and subsidy distortions, monopoly and government regulation, and a host of other realistic frictions and imperfections. These omissions are on the whole likely to produce underestimates of the true costs of a nuclear phaseout. On the other hand, the model clearly cannot see future technological advances, such as improvements in fossil-fuel or transportation technologies, possible developments that would lower the delivered prices of natural gas, or breakthroughs in renewable technologies, and these omissions in the model are likely to lead to overestimates of the costs. We will return to a discussion of some of these realistic features in the final section of this study, but it must be emphasized that, intricate as the model may appear, it is a triviality compared with the actual complexities of the real economy.

BASELINE SIMULATIONS

To estimate the impact of different policies, it is useful to begin with a *baseline scenario*. The baseline is defined here as "business as usual" or "continuation of past trends." For these purposes, the baseline assumes that Sweden builds no new hydroelectric power, that there are no binding new environmental policies (such as constraints on total national emissions of CO_2 or SO_2), and that nuclear power production continues at current levels for the economic lifetime of existing plants. I emphasize that the baseline scenario has no normative significance; it might represent wise or foolish policies. The purpose of defining a baseline is to establish a point of comparison for alternative policies or shocks. The calculation of the impacts of policies—which are measured as the differences between alternative policies—would be *exactly* the same if another set of nuclear policies and environmental assumptions were taken as the baselines.

Table 1 shows the levels and growth of the major variables in the baseline calculation. The major feature of the baseline run is that electricity prices are projected to rise sharply over the coming years as Sweden moves to new and more costly sources of supply. We estimate that capacity is sufficiently expensive that prices must rise considerably above current levels before it is economical to increase capacity to any significant extent. It is calculated that new generation will be with combined-cycle gas turbines, which are projected to be the least expensive source of electricity generation under existing environmental policies. Under the baseline projection, both SO_2 and CO_2 emissions rise as fossil-fuel use increases in the coming years in both the electricity and nonelectricity sectors.

Table 1. Summary Statistics from Baseline Run of SEEP.

	1994	*2000*	*2010*
Energy Demand (tWh)	570.1	612.3	706.4
Specific electric	75.0	78.1	85.0
Transportation	96.4	110.0	136.9
Other	398.7	424.2	484.5
Sources of Electricity Production (tWh)	138.0	138.0	142.0
Nuclear	70.2	70.2	70.2
Hydroelectric	67.8	67.8	67.8
Other existing	0.0	0.0	4.0
New capacity	0.0	0.0	0.0
Environmental indexes			
CO_2 emissions (millions of tons)	50.5	58.3	76.1
SO_2 emissions (thousands of tons)	76.5	88.3	112.7
Energy prices (U.S. cents per kWh)			
Electricity			
Bus bar	2.7	3.3	4.4
Industrial	5.1	5.7	6.8
Residential	10.7	11.3	12.4
Transportation	11.8	11.8	11.8
Other	3.8	3.9	4.0
Externality prices			
CO_2 (U.S. dollars per ton CO_2)	0	0	0
SO_2 (U.S. dollars per ton SO_2)	0	0	0
Externality limits			
CO_2 (millions of tons CO_2)	None	None	None
SO_2 (thousands of tons)	None	None	None
Energy-Output Ratio (1994 = 100)			
Electricity-GDP	100.0	88.95	75.5
Energy-GDP	100.0	95.53	90.9

Toward the bottom, Table 1 shows the projections for the energy-GDP ratio and the electricity-GDP ratio over the coming years. According to the projections, the electricity-GDP ratio is projected to shrink considerably as electricity prices rise in the coming years. The energy-GDP ratio is also projected to decline but at a slower rate than the electricity-GDP ratio. If these projections are accurate, the next fifteen years would see a sharp reversal of the unique feature of recent Swedish economic growth by sharply reducing the energy intensity, and particularly the electricity intensity, of the economy.

This projection should not be taken as the last word on future trends. It is put forth mainly as a reasonable one that is internally consistent and that presents a framework within which we can explore the trade-offs among energy policy, economic policy, and the environment. In the sections that follow, we will use the model to explore alternative scenarios as well as the dilemmas that Sweden faces in the years ahead.

ENDNOTES

[1]William D. Nordhaus, Yale University, Department of Economics, 28 Hillhouse Avenue, New Haven, Connecticut, USA 06511; telephone number in the U.S. is 203-432-3587 (fax 203-432-5779); e-mail is nordhaus@ econ.yale.edu.

[2]Anthony Brooke, David Kendrick, and Alexander Meeraus, *GAMS: A User's Guide,* The Scientific Press, Redwood City, Calif., 1988.

[3]Government of Sweden, Ministry of Finance, *The Medium Term Survey of the Swedish Economy*, Stockholm, 1995, pp. 40, 223.

[4]This specification of energy demand was discussed and derived in William D. Nordhaus, "The Demand for Energy: An International Perspective," in William Nordhaus, ed., Chapter 13, *International Studies of the Demand for Energy,* North-Holland Publishing Company, Amsterdam, 1977. It was used in an energy model similar to the SEEP model in William D. Nordhaus, *The Efficient Use of Energy Resources,* Yale University Press, New Haven, Conn., USA, 1979.

[5]This segment is modeled using a constant elasticity of substitution production function that is calibrated to 1994 prices and inputs.

[6]Some readers have suggested that the electricity demand projections may be high (1) because a significant part of electricity use is displaceable into other fuels and (2) because the electricity market is largely saturated. The first point is likely to be correct but is incorporated in the model because these uses are "general electric" uses that have high substitution possibilities. The second point is a genuine difference of viewpoint because, in my view, saturation is unlikely to occur as new and emerging technologies increase electricity demand.

6

Nuclear Matters

This chapter undertakes an evaluation of nuclear power options for Sweden, and contains the major assumptions and results of the analysis. I begin with a descriptive section in which I analyze the performance of the Swedish nuclear power industry. The first section examines both the economics of Swedish nuclear power as well as the health and safety issues involved. The most controversial issues here involve the potential for serious accidents and the concerns about long-term storage of radioactive waste. Putting the different approaches together, I then estimate the costs of additional capacity in different fuel cycles—nuclear, gas, and coal.

I then turn to the detailed analysis of different scenarios for nuclear power. Using the Swedish Energy and Environmental Policy (SEEP) model just discussed, I estimate the costs of alternative phaseout proposals. One central case to analyze is the one endorsed by the 1980 nuclear referendum in which all nuclear power is phased out in 2010. Three other proposals are to accelerate the phaseout so that it takes place linearly between 2000 and 2010; a phaseout occurring when nuclear power plants reach the end of their economic lifetimes; and a phaseout of only two reactors. I will show the impacts of different phaseout proposals on prices, electricity demands, the overall price level, and economic welfare in Sweden. In addition, we will attempt to determine the extent to which the resulting rise in electricity prices may affect individual industries, with particular attention on the export-intensive industries.

ECONOMIC ASPECTS OF SWEDISH NUCLEAR POWER

I begin with an analysis of the performance of nuclear power in Sweden. It is commonly believed that the nuclear power industry in Sweden is a "success story."[1] This view seems to be based on the fact that there have been no major accidents in Sweden, that the power from the nuclear power sector is produced at a reasonable cost, and that there are not the interminable disputes about siting, cost overruns, or stranded assets that are seen in other countries, particularly in the United States, Britain, and Germany.

To a certain extent, the aura of success comes from the lack of confrontation about siting new plants and facing the controversies about whether alternative technologies are cleaner, more beneficial, and more sustainable. This lack of confrontation is clearly a by-product of the nuclear referendum, which has up to now stilled debate because there was no live opportunity to expand the industry; indeed, it was even unlawful *to plan* for new nuclear power plants.

A second feature of the success of the Swedish nuclear power industry has been the skill and good luck of avoiding any major fiascoes like the U.S. Three Mile Island (TMI) accident. TMI was in effect a multibillion-dollar advertisement campaign against nuclear power, even though the ultimate external economic impacts were minimal (a point to which I return below). I also discuss below the issues of routine health and safety of Swedish power.

The final issue concerns the cost of Swedish nuclear power, to which we now turn. As we showed above, the production costs of nuclear power (particularly the avoidable costs) are absolutely central to the analysis of the economic effects of a nuclear phaseout. There is a wide array of data on production costs that we have examined for the current study.

Nuclear Economic Costs

I begin with an analysis of the total cost for the Ringhals nuclear operation. The Ringhals complex is the largest of the plants and is reasonably representative of the others. It includes one boiling-water reactor (BWR) built in 1976 and three pressurized-water reactors (PWRs) built from 1975 through 1983. The total capacity of Ringhals in 1994 comprised 35% of total Swedish nuclear capacity. Cost data from Ringhals are provided in the annual reports. From the 1992 and 1993 annual reports, we obtain the estimates of the production costs of Ringhals nuclear power shown in Part A of Table 1.

Table 1. Accounting Costs of Production for Ringhals Nuclear Power Stations.

Cost source	*Production cost (cents /kWh)*
A. Average Production Cost in Ringhals Plant, 1987–93[a]	
Operating cost	0.7
Total cost	2.1
B. Accounting Costs for 1993[a]	
Operations and maintenance	0.9
Nuclear fuel	0.4
Charges and taxes	0.3
Capital	0.7
Total cost	2.3

[a]In current prices.

Source: Ringhals Annual Report, 1991, 1992, 1993.

In addition, *Annual Report 1993* provides a more detailed breakdown of the production costs for Ringhals, as is shown in bottom part of Table 1. We have attempted to reproduce these estimates from income-statement data provided by Ringhals staff. Table 2 shows three different concepts of costs that will be useful for the purposes of estimating the costs of a nuclear phaseout. For these, we have calculated the accounting cost of production from Ringhals to be 1.7 cents per kilowatt-hour (kWh), compared with the 2.1 to 2.3 cents per kWh from the annual report. The differences are probably due to the different treatment of nonenergy expenses.

More interesting from the point of view of the current study is the estimate of incremental or *avoidable cost* (those costs that would be avoided if the plants were to be shut down). The estimated avoidable costs include operations and maintenance, taxes, fuel, provision for future costs (for decommissioning and waste storage), and other costs. (Note that some of these are private costs that may not translate into social costs.) I estimate from these data that incremental or avoidable costs were approximately 1.7 cents per kWh in 1993 inclusive of the fee for waste disposal and decommissioning and 1.4 cents per kWh exclusive of the waste fee. These are slightly higher than the estimates in Table 1.

Table 2. Alternative Cost Concepts for Nuclear Power from Ringhals, 1993.

	Total operations (millions of 1993 dollars)	*Energy operations (millions of 1993 dollars)*	*Costs per kWh (cents per kWh, 1993 prices)*
Production (tWh)		20	
Revenues	614	579	2.94
Expenses	345	333	1.69
O&M	176	166	0.84
Fuel	69	69	0.35
Taxes	6	5	0.03
Future costs	61	61	0.31
Other	34	32	0.16
Depreciation	76	72	0.36
Financial items	59	56	0.28
Total costs	480	460	2.34
Net income	134	119	0.60
Estimated future investments	64	61	0.31
Fixed assets in operation	754	712	0.36
Cost estimates (cents per kWh)			
Incremental costs			1.69
Incremental costs plus capital at 10%			2.00
Incremental costs plus ongoing capital costs			2.05

Note: Table shows estimated costs of production using three different concepts. Incremental costs exclude capital charges. Incremental costs plus capital costs at 10 percent capital charge is a reasonable estimate of the market-based cost of capital for comparable investments. The last line includes the average ongoing capital costs from Ringhals investment plans for the period 1994–2004.

A final comparison would include capital costs, shown in the last two lines of Table 2. These are calculated in two different ways. The first imputes a real shadow price of capital of 10% per year on undepreciated capital. The second recognizes that maintaining the plant requires considerable ongoing capital costs (for refitting, new safety requirements, and so forth). These two estimates give total costs of between 2.0 and 2.1 cents per kWh. It is instructive to note

Table 3. Operating Costs of Major Swedish Nuclear Power Plants, 1992–94 (average operating costs in cents per kWh, 1995 prices).

Reactor complex	*1985–94*	*1992–94*
Borseback	1.9	2.0
Forsmark	1.7	1.4
Oskarshamn	1.8	1.8
Ringhals	na	1.6
Weighted average[a]	na	1.6

Note: These show the estimated operating costs, or total costs less depreciation and interest payments. They exceed social avoidable costs because they include certain taxes and fees that cover storage and waste treatment costs that would take place even with a nuclear shutdown. They may underestimate costs because they exclude required ongoing capital expenditures. Data are translated into 1995 prices using the U.S. GNP deflator.

na = not available

[a]Weighted by 1994 production.

Source: Unpublished worksheets provided to the Swedish Energy Commission, 1995.

that the ongoing capital costs of the Ringhals establishment average $64 million per year, which is about $18 per kW of capacity.

A final source on operating costs is unpublished data on costs developed for the 1995 Swedish Energy Commission directed to examine the nuclear phaseout question. These data analyze the accounting records of the four major nuclear power complexes and are summarized in Table 3. The overall findings of this study are that the operating costs for the last three years ranged from 1.4 to 2.0 cents per kWh (in 1995 prices). Using 1994 production data, I estimate the weighted average operating cost of all plants to be 1.6 cents per kWh. This source also provides total (average) production costs, which in 1994 ranged from a low of 1.9 cents per kWh at Ringhals to a high of 3.2 cents per kWh at Oskarshamn.

From these sources, it appears reasonable to conclude that total production costs for nuclear power in recent years have averaged around 2.6 cents per kWh (in 1995 prices), while marginal production costs (inclusive of charges and taxes but exclusive of past and future capital costs and decommissioning and waste disposal fees) were around 1.7 cents per kWh. During the 1987–1993 period, the pretax industrial price averaged 4 cents per kWh. Assuming that transmission and distribution costs were about 1.4 cents per kWh, the two sets of data are reasonably consistent.

How does the performance of the Swedish nuclear plants compare with that of other countries? Again, we have no comprehensive data on other countries, but we can compare the Swedish performance with those countries that have a high share of nuclear power as well as with the costs of new nuclear additions. We noted above the estimates of pretax industrial electricity prices in a number of industrial countries. The two other European countries with large nuclear shares are France and Belgium, with nuclear shares in generation of 78% and 59% in 1993. Industrial electricity prices in both countries are currently well above those in Sweden. These comparisons may not be entirely appropriate, however, because the Swedish production contains a high share of hydroelectric power. The estimates of the cost of nuclear power given above, which are close to the average for the estimated industrial generation costs, suggest that the comparison may not be seriously biased. Another reason for the difference is that Sweden has undergone a major currency depreciation relative to the hard-currency countries of northern Europe. This will lead prices based on historical costs—as do electricity prices, which are regulated on the basis of historical costs—to lag behind other prices.

To a certain extent, therefore, electricity prices in Sweden may be depressed relative to their true social costs. Nonetheless, the fact that France is generally thought to be an efficient producer with perhaps the most successful nuclear industry in the world—along with the fact that Swedish industrial power prices are below those in France—suggests that Swedish nuclear power is in fact relatively efficient.

A second comparison is of the current costs of production of Swedish nuclear power with the estimated costs of new nuclear power plants. I present below estimated costs of new generation from a survey of experience of different countries by the International Energy Agency (IEA). The IEA also provides estimates of the total cost of generation from different sources. According to compilation shown in Table 14 on page 111, the levelized cost of nuclear power plants currently in planning is 4.8 cents per kWh at a 5% discount rate and 7.1 cents per kWh at a 10% annual discount rate. Clearly, the current average cost of operating Swedish nuclear plants (around 2.6 cents per kWh) is well below the estimated costs of new nuclear generation in the major industrial countries.[2]

In summary, there are good accounting data on the economic performance of Swedish nuclear power, but comparisons with other countries are difficult. The existing data suggest that the production costs in Sweden are comparable to or below the nuclear power costs in other countries and are well below the costs of new nuclear power production.

Nuclear Capacity

An important question concerns the capacity and production of nuclear power plants over the coming years. Clearly, if nuclear power continues to have high levels of availability and low cost, then a shutdown will be relatively expensive. On the other hand, if the productive capacity declines as the plants age, then shutdown costs will be relatively lower.

In fact, nuclear capacity has been creeping up over the last decade as the plant operations have been fine-tuned. The average net generation over the 1990–1994 period averages 65 terawatt-hours (tWh) per year, while for 1994 net generation from nuclear plants was 70.2 tWh. There is clearly some uncertainty about the future availability of nuclear power over the next two decades. The SEEP model follows the convention of calibrating to 1994 data, so the baseline runs set capacity at the production for that year of 70.2 tWh. To the extent that actual availability is greater or lesser than that, the estimates of the cost of a total nuclear phaseout will be proportionally changed.[3]

Another way of seeing this point concerns the *cost* of keeping existing reactors operational. It is possible that the operating costs will rise if the costs of replacing equipment or maintaining high safety standards increases sharply over time. To the extent that such maintenance costs rise sharply, the costs of a nuclear phaseout would be reduced in a fashion parallel to the impact if the nuclear capacity were reduced because some plants became nonoperational.

THE TAIL END OF THE NUCLEAR CYCLE

An examination of the future of Swedish nuclear power requires an analysis of the tail end of the nuclear power cycle. The questions involved here are the useful or economic lifetime of nuclear power plants, the costs of decommissioning the plants when they are put to rest, and the costs of disposing of the wastes. Because the industry is still relatively young, there are few reliable observations on any of these issues, but they clearly have a significant impact upon the economics of a phaseout. We review the salient questions in this section.

Lifetimes of Plants

The first question involves the useful or economic lifetime of nuclear power plants. This issue is clouded because it is not simply a technical issue (like the lifetime of a lightbulb) but depends largely on the rigor and quality of maintenance and replacement, on the costs of operation and alternative power sources, and on the regulatory framework. Conventional power plants are akin to houses or cathedrals, which outside of wartime can have their lives indefinitely extended, so that after a few decades very little of the original equipment remains. We do not have sufficient experience with nuclear power plants to know whether their lifetimes can also be indefinitely extended. The issue is also complicated because the regulatory bias against building new plants of any kind, and nuclear power plants in particular, gives a strong incentive to extend the lifetime of existing plants beyond the point at which it would be optimal to replace the plant, were old and new plants to be treated neutrally.

The concept of economic lifetime is quite different from a technical definition or a tax lifetime. Some confusion arises because of the distinction between economic lifetimes and accounting or tax lifetimes. The original Swedish nuclear power plants often had accounting lifetimes of twenty-five years. From this, it was easy (but incorrect) to conclude that they would be obsolete or uneconomical in twenty-five years. In fact, firms often employ short tax lifetimes since they are advantageous from a firm's point of view because they increase the capital recovery rate, accounting cost, tax-deductible expenses, and allowable price in most price-regulated systems. But in many cases, tax lifetimes are shorter than would be justified purely on the basis of the economic value of an asset. We see today many assets that are performing profitably long after their book values for tax or accounting purposes are zero.

The appropriate lifetime to examine is the "economic lifetime," which is that point at which the present value of net revenues from continued operations no longer exceeds the present value of the plant's closure. Lifetime projections for most nuclear power plants are considerably longer than the original planning periods. A recent Nuclear Energy Agency survey concluded:

> Many nuclear plants are coming towards the end of the lives originally planned for them. These lives were based on conservative technical assumptions and the adoption of considerable margins of safety, based on the state of knowledge and experience that existed 20 years ago. It is now realized that many of the existing reactors could continue to operate safely for periods significantly in excess of their original design lives.... There appears to be no inherent reason why water reactors can not have their lives extended to well beyond the 40 years that some designers now project.[4]

The actual lifetimes are still a matter of conjecture because the actual experience with older plants is limited and the newer vintages are generally more reliable. Of the power plants in use today around the world, none have lifetimes in excess of forty years, only four are older than thirty years, and seventy-seven are between twenty and twenty-nine years old.[5] A few of the older plants have in fact been retired. The current view in the nuclear industry was exemplified in discussions I held with technical staff at the Ringhals complex, in which I was told that Ringhals managers operate their plant by replacing equipment in such a way that it could operate indefinitely. The practice of sustainable development economics has definitely caught up with the nuclear power industry!

There is clearly no correct answer as to the lifetime of Swedish nuclear power plants. A reasonable baseline assumption, which is consonant with current international practice, is for a lifetime of forty years from the start of operations, which would put the average expected date of retirement at 2020. We can test the sensitivity of the results to this assumption by testing the impact of thirty- and fifty-year lifetimes as well.

Death and Interment for the Nuclear Fuel Cycle

The back end of the nuclear fuel cycle has proven to be the most durable source of controversy and uncertainty. Major issues surround the decommissioning of nuclear facilities and the disposal of nuclear wastes. Sweden has made considerable progress and achieved a high degree of consensus in this area compared with virtually any other country. I review briefly the issues involved in decommissioning and nuclear wastes in this section.

The treatment of funeral expenses for nuclear power stations is somewhat paradoxical. Before the plants are constructed, decommissioning costs should be included as part of the total costs of going down the nuclear path. However, once the plants are in operation, the costs of decommissioning and disposal are no longer avoidable options and the only question is when and how carefully interment takes place. An early phaseout will actually raise the decommissioning costs because this activity is moved closer to the present. In a sense, the decommissioning costs might be better thought of as a cost of the replacement power rather than as a cost of nuclear power.

Decommissioning. At the end of a facility's useful life, the facility is "decommissioned" and removed from service. The analysis of decommissioning usually divides the process into three stages. In Stage 1, the mechanical systems are blocked and sealed. In Stage 2 decommissioning, the containment barrier is reduced to its minimum size, all easily dismantled parts are removed, and the remaining barrier is sealed. Stage 3 decommissioning is the final step in which all materials and equipment whose radioactivity is significantly above natural background levels are removed. The site can then be used for nonnuclear purposes. Most countries plan to undertake Stage 1 and Stage 2 decommissioning shortly after shutting down a nuclear power station, but Stage 3 can be postponed for decades to reduce the cost and to allow radiation levels to decay.[6]

A great deal of study has been made of the feasibility and costs of decommissioning. There has been to date one complete (through Stage 3) decommissioning, that of the first commercial nuclear power plant, the Shippingport Atomic Power Station in Pennsylvania. The total cost was $120 million.[7]

Estimates have been made of the costs of decommissioning Swedish power reactors; the most recent ones date from the 1980s and estimate a cost of $1.24 billion for all operating plants.[8] For the two plants most carefully studied, these represented costs of approximately $215 per kWe (kilowatts of electrical energy). The costs for Swedish decommissioning were somewhat lower than those in other countries, which show an average estimate of about $230 per kWe; this is somewhat surprising because the Swedish costs undertake Stage 1 through Stage 2 decommissioning very quickly.[9] Comparable costs for the countries surveyed range from $115 to $415 for BWRs and PWRs in other countries. If the correct number is $215 per kWe, this can in effect be thought of as the additional cost of replacement power if nuclear power is phased out. The levelized cost of this addition is 0.17 to 0.33 cents per kWh at 5% and 10% discount rates.

Waste Disposal. The other major item of the tail end of the nuclear fuel cycle is waste disposal. This has proven enormously controversial in all countries. Until recently, Sweden was the only country that had reached a technical and political consensus on waste disposal (in 1995, further questions were raised and that consensus may be disintegrating). According to Swedish law, all costs of decommissioning and waste disposal shall be the responsibility of the power producers. For this purpose, they have established the Swedish Nuclear Fuel and Waste Management Company (SKB), which has the responsibility of managing and disposing of Sweden's radioactive wastes. The activities of SKB are financed by a fee levied on nuclear power producers; that fee averaged 0.28 cents per kWh of nuclear power generation in 1993 (in 1995 prices). The breakdown of the prospective uses of these funds is the following:

- Decommissioning: 20%
- Deep storage: 36%
- Interim storage: 15%
- Transportation: 4%
- Reprocessing: 10%
- Other: 15%

SKB has identified and is in the process of establishing a specific site and technique for long-term storage of nuclear wastes. This is a site, close to one of the operating nuclear power plants, approximately five hundred meters deep in the bedrock, with multiple layers of defense against leakage of the radioactive wastes into the groundwater.[10] It remains to be seen whether SKB's plans for disposal will proceed on schedule.

Impacts on Phaseout Costs. As discussed above, the question of decommissioning and waste disposal is important for the nuclear phaseout because an earlier phaseout will *increase* the discounted costs of nuclear power by bringing those costs closer to the present. Table 4 shows the basic data on the current and projected future costs of decommissioning Swedish nuclear power plants. SKB estimates that the levelized cost will be 0.285 cents per kWh (see the second to last row of Part B of Table 4). I estimate that approximately 70% of these are deferrable costs. In terms of totals, I estimate that of the $164 million contributed annually to the tail-end funds administered by SKB, $116 million would be deferred for each year that a nuclear phaseout is deferred. Putting this differently, I estimate that of the 0.285 cents per kWh of disposal costs that is currently embedded in nuclear power costs, 0.197 cents per kWh of that cost would be eliminated if the plants were to remain open indefinitely.

The last part of Table 4 shows two different estimates of the capital cost of decommissioning Swedish nuclear power plants. The first row is derived from the SKB data and shows the likely avoided costs, while the second is the estimate of decommissioning from an earlier study of the Swedish nuclear power industry. These two estimates are remarkably close. For the estimates that are prepared below, I assume that the costs of shutting down the nuclear plants are $215 per kWe centered in the year of the shutdown.

SAFETY OF NUCLEAR POWER

From the outset, the nuclear power debate has centered on the issues of safety, environmental quality, and external impacts. Much experience has been accumulated since the first commercial nuclear reactor opened in 1956 and since the Swedish referendum of 1980. In this section, I review the record on nuclear power safety and the outstanding issues and compare the Swedish performance with those of other countries. Finally, I conclude this section with a summary comparison of the total external costs of different fuel cycles in electricity production.

In analyzing the safety of nuclear power, it is useful to separate the concerns into three categories: routine releases, severe accidents, and waste disposal. Each of these is rife with popular confu-

Table 4. Breakdown of Disposal and Decommissioning Costs (1995 $U.S.).

A. Operating Costs of SKB, 1993

Segment	*[1]* *$ millions*	*[2]* *Cents per kWh*[a]
Research and planning	23.6	0.040
Intermediate storage (CLAB)	10.4	0.018
Low level wastes (SFR)	3.6	0.006
Transportation	2.6	0.004
Fuel services	75.9	0.130
Capital costs	17.9	0.031
Other	7.1	0.012
Total costs	140.9	0.241
Set aside for future[b]	239.4	0.409
Total charges	380.3	0.650

B. Estimated Final Costs

		Levelized costs (cents per kWh)		
Segment	*[3]* *percent*	*[4]* *Current estimate*	*[5]* *Incurrred with phaseout*	*[6]* *Percent of costs incurred with phaseout*
Intermediate storage (CLAB)	15	0.043	0.021	50
Deep repository	36	0.103	0.092	90
Low level wastes (SFR)	5	0.014	0.004	25
Transportation	4	0.011	0.003	25
Reprocessing, terminal	10	0.029	0.026	90
Decommissioning	20	0.057	0.051	90
Other	10	0.029	0.000	0
Total costs	100			69
Cents per kWh		0.285	0.197	
Total, million dollars per year		164	116	

continued on next page

Table 4. Breakdown of Disposal and Decommissioning Costs (1995 $U.S.)—*Continued.*

C. Estimated Total Costs of Discontinuation

	[7] *Total* *($ billions)*	*[8]* *per kWe* *(dollars)*
Capital Costs:		
Discounted annual costs	2.3	223
Direct estimates	2.1	213

[a]Calculated on the basis of 56.8 tWh of generation.

[b]Difference between revenues and expenses plus interest on reserves.

Source: SKB, *Activities 1993*, for columns [1] and [3]. Column [2] calculated by author on basis of actual generation as explained in note (a). Column [4] calculated assuming fee of 0.285 cents per kWh from SKB. Column [5] calculated from columns [6] and [4]. Column [6] assumed by author. Entry 1 in column [7] is the discounted value of total cost in column [4], while the first entry in column [8] divides that by total gross capacity. The second entries in [7] and [8] take the estimate for decommissioning of the Ringhals plant from the NEA study cited below. All values are converted into 1995 $U.S. using the GDP deflator.

sion and scientific controversy, and there are no cut-and-dry answers anywhere. Therefore, to a larger extent than in most parts of this study, the analysis here represents the author's point of view and synthesis of the outstanding issues and data rather than an attempt to determine a scientific consensus where none exists.

Routine Releases

For the most part, nuclear power in Organisation for Economic Co-operation and Development (OECD) countries in general and in Sweden in particular has had minimal to modest routine health and safety costs relative to most of the economy or to other forms of electricity or energy production. The low level of routine occupational and public exposure today is the result of extremely stringent regulation of commercial nuclear power.

There are numerous estimates of the risks of nuclear power. The top half of Table 5 shows estimates of the risk from nuclear power

Table 5. Estimated Fatality Rate in Nuclear Power Production and Other Sectors (1995 $U.S.).

	Fatalities	*Value of Production ($ billions)*	*Fatality (fatalities per $ billion of output)*
Nuclear power			
Entire production cycle[a]	50	26	1.90
Routine operation of Swedish reactors[b]	0.03	2.8	0.01
United States, 1992			
Agriculture	1292	109	11.86
Mining	215	92	2.34
Manufacturing	646	1026	0.63
Transportation and public utilities	1292	506	2.55
Services (including FIRE)	1400	2130	0.66
Total economy (GDP)	9155	5723	1.60

Note: Fatality rate estimated to be .025 fatalities per man-Sievert-year (see discussion in sources below). Estimated fatality rate includes power-plant radiation only.

[a]Calculated at rate of 100 gWe-years at a price of 3 U.S. cents per kWh.

[b]Value of production calculated as 70 tWh per year at a pretax market value of 4 cents per kWh.

Sources: Estimates of nuclear power from Nuclear Energy Agency, *Broad Economic Impact of Nuclear Power*, OECD, Paris, 1994. Data for United States from U.S. Commerce Department, *Statistical Abstract of the United States, 1994*, Washington, D.C., 1994. Estimate of exposure rate for Sweden from Table 20. Estimate of health impact of radiation from U.S. National Academy of Sciences, *The Effects on Populations of Exposure to Low Levels of Ionizing Radiation*, Washington, D.C., National Academy Press, 1980.

operations. The first entry is the estimated fatality rate for the entire nuclear power production cycle according to a recent international workshop to be discussed shortly. These estimates include mining, which is the riskiest component, and other parts of the cycle. The second entry shows the estimated fatality rate from radiation exposure of routine operations of the reactor. As is clear, the major risks come outside the routine operation of the reactor itself.

The second half of Table 5 compares the risks from nuclear power with other occupational hazards in the other sectors of the U.S. economy. These data are not completely comparable because the data for nuclear power include external costs, while those for other sectors are entirely occupational fatality data. The impacts of nuclear power production are on the same order as the occupational hazard rate for the entire U.S. economy—about two fatalities per $1 billion of output. The fatality rate is approximately one-tenth of the rate in that bucolic occupation, farming, and is approximately three times more hazardous to the national health than are services like education or insurance. On the other hand, the hazards from the routine releases of the reactors themselves are about one-hundredth as large as those from average economic activity.

The numbers in Table 5 are extremely impressionistic and are shown only to provide a rough comparison of health and safety impacts in different industries. The estimates of external fatalities of production outside the energy sector are probably understated by an order of magnitude because of the omission of the impacts of air pollution and similar factors, so a more complete accounting would probably show nuclear power production to be significantly *less* risky than the balance of the economy. We will provide below more careful comparisons of nuclear power and other sources of electricity generation. As later sections indicate, nuclear power production is much less harmful in terms of societal fatalities than are most fossil-fuel electricity production cycles, particularly that of coal burning.

How does Sweden compare with other countries in terms of routine operations? Data on performance of commercial nuclear power operations are collected by WANO (the World Association of Nuclear Operators). The WANO data pertain to a wide variety of quantifiable attributes of nuclear power in different countries.

Table 6 compares the performance of Swedish nuclear power plants with those of other WANO countries for three major indicators (see the accompanying footnote for precise definitions). The number of unplanned major shutdowns or "scrams" was comparable for Sweden and other countries. This indicator is not dispositive, however, because scrams can occur either because of poor equipment or maintenance or because the tolerance levels for scrams are demanding.

Table 6. Performance Characteristics of Nuclear Power, Sweden and WANO Countries (per reactor).

Attribute[a]	*All countries, 1990–92*	*Sweden, 1990–92*
Reactor scrams (per year)	1.1	1.2
Collective radiation exposure (person-Sieverts)		
PWRs	1.7	1.0
BWRs	2.6	1.6
Energy availability (%)	77	81

[a]Unplanned scrams are automatic shutdowns of the reactor that were not anticipated parts of a planned test. Energy availability is the fraction of the period during which energy could have been produced. Collective radiation exposure (person-Sieverts) is the total whole-body dose measured in person-Sieverts per reactor-year received by all on-site personnel during the period. Conventional estimates for low doses are that 10,000 person-Sieverts will produce 125–1000 lifetime cancers. The data are sometimes separated between PWR (pressurized-water reactors) and BWR (boiling-water reactors).

Source: World Association of Nuclear Operators, *1992 Performance Indicators*, WANO and Kärnkraftsäkerhet och Utbildning AB, *Operating Experience in Swedish Nuclear Power Plants, 1993*, Stockholm.

The level of radiation exposure was lower in Swedish plants than the average of other countries. These data are unambiguous in principle, but it is likely that rigorously monitored plants will have a higher reporting ratio than carelessly monitored or regulated systems. In addition, there are major differences across vintages of plants, with the oldest vintage of Swedish plants having exposures five times those in the youngest vintage. Finally, it should be emphasized that the total exposure is very modest in comparison to other industries as was shown in Table 5.

The final attribute is the plant availability. A high availability indicates effective management and absence of extended unplanned shutdowns. Swedish nuclear power stations have had high availability relative to other countries, which reflects the high standards of engineering in the Swedish plants. Since 1992, particularly in the older BWR plants, problems with leakages and cracks in containment vessels have been detected and performance in these systems has been worse than the Swedish and international average. Whether recent poor availability foreshadows incipient senility of these plants or is just a bad cold remains to be seen.[11]

Overall, the routine performance of Swedish nuclear power has been superior by international standards and in comparison to other risky industries.

Catastrophic Accidents

Whereas routine health and safety impacts are by their nature relatively easily estimated given the extensive experience with nuclear power to date, estimating the risks from low-probability, high-consequence accidents has been extremely difficult and highly controversial. Early studies of the probability of major accidents include the well-known Rasmussen Report.[12] This landmark study estimated the probabilities of accidents of different severity in light-water reactors. One of the conclusions of the Rasmussen Report was a set of estimates of the probabilities of accidents due to human-caused events according to severity; a representative set of estimates is shown in Table 7.

These early estimates were thrown into doubt by the Three Mile Island accident in 1979 and by the catastrophic Chernobyl accident in 1986. Retrospective analyses of the TMI accident indicate that the total external costs were on the order of $30 million with estimates of fatalities on the order of 0.7.[13] Taking one fatality as the benchmark number for the health impact of TMI, an accident of this magnitude was estimated by the Rasmussen Report to occur once in every 31,000 reactor-years in the most likely case. At that time, the total number of reactor-years was four hundred, so it

Table 7. Probability of Accidents of Differing Levels of Severity According to Rasmussen Report.

Severity (fatalities per million reactor-years)	*Frequency (per million reactor-years)*		
	Optimistic	*Most-likely*	*Pessimistic*
At least 1	6.3	32	160
At least 10	4.0	20	100
At least 100	0.5	2.5	13
At least 1000	0.006	0.032	0.16

Source: Reactor Safety Study, op. cit., pp. 88 and 90.

might appear that the Rasmussen estimates significantly understated the probability of major accidents.

More recent estimates have confirmed the general conclusion that the risk of core meltdowns in modern light-water reactors (LWRs) is low. The target design in the United Kingdom is that the probability of a core melt from all causes should be less than one in a million. A recent study by the Dutch government concluded that the probability of a core melt in future LWRs will be between one in ten million and one in one hundred million per year.[14]

For Sweden, a government decision in 1986 stated that in the case of an accident involving severe core melt, releases should be limited to a maximum of 0.1% of the core content of cesium 134 and 137; by contrast, between 20% and 40% of the core context of these isotopes were released in the Chernobyl accident. The Swedish nuclear regulatory agency, SKI, estimates that current operations and hardware will ensure that the probability of such an accident should be less than one in one hundred thousand, which is about one-tenth the frequency estimated by the Rasmussen Report.[15]

These estimates are normative rather than predictive. Moreover, estimating accident risks is a difficult task. Another approach to estimating the probability of a severe accident is to ask what is the most likely distribution of accidents given the actual historical occurrence of accidents. For this purpose, we consider the risk of accident in light-water reactors of standard European or American design; for this purpose, we exclude Russian, experimental, or breeder reactors. We would clearly want an entirely different analysis for graphite reactors, such as the kind operated at Chernobyl.

Assume for this calculation that the TMI accident is the most severe accident to date and that that accident had the equivalent of one off-site fatality. Table 8 shows the empirical or historical estimate of the maximum likelihood estimate of the probability of severe accidents given the number of reactor-years of history and the fact that the TMI accident is the most severe accident on record.[16]

The results using the historical data indicate that the Rasmussen Report significantly underestimated the risk of a TMI-sized accident at that time, when there were but four hundred reactor-years of experience. There has been a great deal of further experience since TMI, and the accumulated reactor-years today is approaching five thousand. Given five thousand reactor-years, the maximum

Table 8. Probability of Alternative Accidents Given Three Mile Island.

Length of historical record (reactor years)	*Maximum likelihood frequency (per million reactor-years) of:*	
	TMI accident	*Severe accident*[a]
400	2500	2.5
5,000	200	0.2
25,000	40	0.04
Memo: Rasmussen Report estimate	32	0.032

[a]A severe accident is one that has at least 1000 off-site fatalities.

Note: This table uses a Bayesian analysis to calculate the probability of an accident with at least one off-site fatality (for the TMI) and at least 1000 off-site fatalities (for the severe accident). The top three rows show the maximum likelihood estimate assuming that the given number of reactor-years of historical experience has witnessed no accident more severe than TMI.

Source: See text.

likelihood estimate of a TMI-sized accident is two hundred per million reactor-years. Moreover, assuming that the Rasmussen Report has correctly calculated the ratio of probabilities of severe (one thousand fatalities or more) to TMI-sized accidents, the maximum likelihood estimate of a severe accident is today about 0.2 per million reactor-years. This calculation is still far above the Rasmussen Report estimate of 0.032 per million reactor-years for severe accidents. On the other hand, this empirical estimate is not far from the design frequencies—between one in one million and one in one hundred million for a core melt—that were discussed above. Using the historical technique, we see that we would need another twenty thousand reactor-years of experience (approximately sixty years at current capacities) to lower the historically based probabilities to the Rasmussen estimates—assuming that there are no accidents worse than TMI.

At this point, combining the historical data and the theoretical risk assessments, I judge that a reasonable estimate of the risk of a severe accident (defined as either a core melt or an accident with more than one thousand fatalities) lies between one in a million reactor-years and one in one hundred million reactor-years. The interesting point is that with the extensive experience with light-water reactors, the historical accident frequency lies within the range of the theoretical risk analyses.

It is difficult to make any firm conclusions about the likelihood of serious to catastrophic accidents. The history to date has been encouraging for well-managed systems like those currently operated in Sweden. The fact that LWRs have five thousand reactor-years of experience with no accidents graver than the Three Mile Island experience lends credence to the view that another ten to forty years of operation of nuclear power in Sweden poses relatively low risks. Using the maximum likelihood estimates in Table 8, we calculate that if Sweden operates its current reactors for their projected economic lifetime of another twenty-five years, the probability of a TMI-sized accident or worse is around 6% and the probability of a severe accident is around 0.005%. To the extent that Swedish nuclear power is better (or more badly) managed than average, these figures would be lowered (or raised). In addition, if experience has led to improved management and regulation of nuclear facilities, then the probabilities of accidents have declined over time and these probabilities might thereby be overstated.

While we are relatively sanguine about the safety of Sweden's nuclear power industry, it must be emphasized that there are inherent dangers in the nuclear power technology. One set of dangers lies in inherently unsafe technologies, such as the graphite reactors used at Chernobyl and in many reactors in Russia and eastern Europe. The difficulties here arise because accidents can lead to enormous releases of radioactive material. A second set of dangers occurs when nuclear reactors are badly managed or regulated, where even well-designed systems can lead to serious accidents. Only through the combination of good reactor design and vigilant oversight and management can we be relatively secure that the inherent dangers in nuclear power are reduced to socially acceptable levels.

Long-Term Storage

At the tail end of the nuclear power life cycle, the final problem is the disposal of the nuclear wastes, particularly the high-level and long-lived wastes arising from spent fuel reprocessing. In Sweden's case, a nuclear shutdown will require finding a retirement home for the unspent fuel as well, in effect extending Sweden's welfare state to nuclear fuel.

The approach to waste disposal favored by most experts is geological disposal. This would encase the wastes in canisters and then deposit them in mined cavities deep underground. After one thousand years, the remaining radiation is at a level of a body of natural uranium ore. Many experts believe that the risks from geological disposal are less than the current storage at reactor sites.

Notwithstanding the expert consensus, the prevailing suspicions of nuclear experts has made resolving the long-term storage problem a thorny issue for most countries for a number of reasons. To begin with, the long lifetimes of the high-level wastes imply that a solution will have to protect the public health and safety for millennia, whereas the durability of existing political arrangements is often measured in decades or, in rare cases, for a few centuries. In addition, experts are divided on the issue, with technical people in major countries often unable to agree on the best or even a reasonable solution. Finally, even when a technical solution is agreed on, most democracies are plagued by the NIMBY (not in my backyard) syndrome, where people continue to fight the location of nuclear waste storage sites in their localities, even when those projects bring significant infusions of jobs and incomes.

We have discussed the issues involved in decommissioning and long-term storage. These are designed for Sweden so that the health and safety impacts are negligible, and the estimated costs are included in the life-cycle costs of different fuel cycles analyzed below. Because long-term storage is currently only on the drawing boards, however, we have no historical experience on which to draw to test the reliability of the estimated safety.

In any case, at this stage the issue is moot. Once the nuclear plants are built and operating—once the Faustian nuclear bargain has been struck, so to speak—there is no exit from the road to finding a solution for long-term storage. To a certain extent, the storage issue works against the logic of a nuclear shutdown. The more rapid the shutdown, the sooner will Sweden be forced to decide where it will house its nuclear wastes. To the extent that two decades of experience may lead to improved scientific and engineering knowledge about how safely to dispose of nuclear wastes, this suggests that postponing the decision will enhance the safety of the disposal, and a slower phaseout of the nuclear industry would in fact be a de facto postponement of the long-term storage problem. But if current estimates of the safety of the current proposals for long-term storage are close to accurate, there is but a minimal environmental difference between a phaseout in 2010 and waiting until the current generation of plants dies its natural economic death.

Nuclear Proliferation

A further question sometimes raised, the relationship between nuclear power and the proliferation of nuclear weapons, is beyond the scope of this study. A few comments on the issue will be useful, however. The concern about nuclear proliferation arises both because of the enormous volumes of nuclear material in nuclear power reactors and because of the low levels of protection in civilian reactors compared with military nuclear arsenals. At the same time, it must be noted that the material available in most nuclear power plants is unsuitable for bomb making without expensive and sometimes dangerous further chemical treatment. A final concern is the spread of dangerous nuclear material outside of countries where it currently resides into bandit states like North Korea or Iraq.

A perspective on this issue is found in a 1992 study by David Fischer.[17] He notes that many countries that have robust nuclear power programs (such as South Korea, Sweden, and Japan) have shown no inclination to develop nuclear weapons. Other countries that are developing nuclear weapons (such as India and Pakistan) are facing obstacles to developing nuclear weapons, not because of the regulation of nuclear power plants but primarily because other countries are impeding their development of reprocessing and enrichment plants. Countries such as Brazil, Argentina, and South Africa that have moved away from nuclear weapons did so because of political changes rather than technical developments.

Moreover, the two governments that have most recently pursued nuclear weapons—Iraq and North Korea—have taken the direct route by building dedicated facilities rather than by diversion from a civilian nuclear power station. The direct route was the one followed by each of the five recognized nuclear states as well as by Israel. The only example where a government appeared to be on the route to using a nuclear-generating station for bomb-making purposes was Iraq's use of the Tammuz 1 station. When Israel destroyed the station, Iraq then switched to uranium enrichment. Fischer concludes his survey of nonproliferation as follows:

> [It is] difficult to discern any direct link except in the negative sense between civilian nuclear power and nuclear weapons. ... [As] Saddam Hussein and Kim Il Sung have shown once again, the decision to acquire nuclear weapons or to keep the option open is political, not economic. We now surely have enough evidence to show that if a nation that has reached an intermediate level of industrial development believes that nuclear weapons or the option to make them is essential for its own security it will in time be able, by hook or by crook, to gain mastery over a sensitive nuclear technology. We can lengthen that time by external controls, but to stop the spread of nuclear weapons and to reverse the race between those that have them, we must mitigate the security concerns that stimulated the spread and kept the race going.[18]

It should be noted that this view is disputed by other experts on nuclear proliferation. For example, Theodore Taylor argues that the vast quantities of nuclear materials provide many opportunities for danger. He emphasizes the dangers of bombing or sabotage of nuclear power plants and recommends a rapid phaseout of nuclear power.[19]

ESTIMATES OF EXTERNAL COSTS FROM DIFFERENT FUEL CYCLES

We have discussed the external costs of nuclear power in isolation, without comparing these costs with the alternatives. If Sweden decides to phase out nuclear power, it will need to find alternative sources of power. Given other constraints, the most likely sources of domestic generation would be natural gas–fired or coal-fired plants. How do the environmental costs of these alternative fuel cycles compare with nuclear power?

There have been numerous studies comparing fuel-cycle externalities, and Table 9 presents the results of a recent international workshop on the subject. The figures on fatalities are most easily compared across fuel cycles. By contrast, the nonfatal incidents contain a wide variety of health effects from emphysema to nose irritation and are therefore relatively less useful in such an aggregated form. In addition, it might be questioned whether the occupational hazards are really external if workers are compensated for them. The basic conclusion of this table, which is replicated in virtually every study of the health effects of different fuel cycles, is that nuclear power and natural gas contain relatively small societal health and safety risks, while coal is a dirty and nasty fuel with significant external costs as well as occupational hazards.

Table 10 collects the results from the different tables to estimate the total environmental, health, and safety costs of the different fuel cycles. These value fatalities or fatalities prevented using recent evidence on the willingness to pay for fatalities and use a central figure of $4 million per life saved. The nonfatal incidence is grossed up from the fatalities using recent evidence on the ratio of total health

Table 9. Comparative Risk by Electricity Production by Fuel Cycle (accidents and diseases per gigawatt-electric-year).

	Occupational hazards		*Public (off-site) hazards*	
Fuel Source	*Fatal*	*Nonfatal*	*Fatal*	*Nonfatal*
Coal	0.2–4.3	63	2.1–7.0	2018
Oil	0.2–1.4	30	2.0–6.1	2000
Gas	0.1–1.0	15	0.2–0.4	15
Nuclear (LWR)	0.1–0.9	15	.006–0.2	16

Note: This table shows the estimated risk to workers in the industry and the external costs to the public of producing 1 gigawatt-electric-year of power from different sources. The data include extraction, processing, transport, and operation. The risk of severe accidents is excluded from all cycles. Note that some costs may be internalized to the extent that worker hazards are compensated through higher wages or in other ways.

Source: Data on fatalities from A. F. Fritzsche as adopted by the Senior Expert seminar in Helsinki. Nonfatal incidence from W. Paskievici. Both are as cited in Nuclear Energy Agency, *Broad Economic Impact of Nuclear Power*, op. cit., p. 73f.

Table 10. Estimated External Costs from Accidents and Injuries by Electricity Fuel Cycle (U.S. cents per kWh, 1991 prices).

Fuel source	*Fatalities*	*Nonfatal accidents*	*Severe accidents*	*Total*	*Item: total costs of accidents as percent of electricity price*[a]
Coal	0.311	0.155	0.0000	0.466	11.6
Oil	0.221	0.111	0.0000	0.332	8.3
Gas	0.039	0.019	0.0000	0.058	1.5
Nuclear (LWR)	0.028	0.014	0.0002	0.042	1.0

[a]Electricity cost at 4 cents per kWh.

Note: This table calculates the cost per kWh of fatalities, accidents, and severe events for each conventional fuel cycle. To calculate the cost of a fatality, it is assumed that the willingness to pay for a statistical life saved is $4 million, following studies of W. Kip Viscusi (see W. Kip Viscusi, "The Value of Risks to Life and Health," *Journal of Economic Literature*, Vol. 31, No. 4, December 1993, pp. 1912–1946). Costs of nonfatal accidents and diseases are assumed to have a value of one-half of the costs of fatalities based on the results of a study by the U.S. Environmental Protection Agency on the economic costs of air pollution (U.S. Environmental Protection Agency, *The Benefits and Costs of the Clean Air Act, 1970 to 1990*, May 3, 1996 Draft, Washington, D.C.). Probabilities of a severe accident are estimated using the historical data from Table 22; for those calculations, it is assumed that the willingness to pay is four times that of a routine accident. It is assumed that there are no severe accidents associated with fossil fuel cycles.

effects to costs of fatalities in studies on the economic effects of air pollution.[20] The estimates of catastrophic or severe accidents are taken from the estimates in Table 8 using five thousand reactor-years of experience and assuming that nonfatal illnesses have a value equal to that of fatal illnesses, while property damage and other effects are equal in value to human health effects.[21]

Aggregating these data in Table 10, we draw three conclusions. First, the nuclear and natural gas fuel cycles pose significantly fewer health and safety risks and costs than do the coal and oil fuel cycles. The difference between the high-cost and the low-cost cycles isapproximately an order of magnitude. Second, the estimates presented here indicate that the total cost from health and safety effects of nuclear power are on the order of 0.04 cents per kWh for the representative nuclear power cycles examined in the underlying studies. Given current electricity prices, this implies that the total health and safety costs associated with nuclear power are around 1% of the total production costs. The external costs of power for coal are significantly higher—around 12% of electricity prices according to Table 10.

Third, the expected value of the costs of severe accidents from nuclear power is estimated to be a tiny fraction of the total health and safety costs of either nuclear power or of other fuel cycles. This suggests that excessive attention may have been given to the severe risks of nuclear power, while insufficient attention has been paid to the more mundane risks, particularly those associated with oil- and coal-burning facilities.

This concludes the discussion of the environmental risks and costs of nuclear power. On the whole, this review indicates that continued operation of Swedish nuclear power plants is likely to impose a relatively small environmental, health, and safety burden on Sweden. This conclusion is particularly the case if a nuclear phaseout is followed by increased generation with either coal-fired or oil-fired electricity generation. Having said this, we must recognize that all such calculations are subject to wide margins of potential error and that the estimation of health and safety effects of energy systems has far to go before we can fully understand their effects. Moreover, those relative risks that look small and acceptable in the aggregate may appear frightening and unacceptable to individuals living near the plants, particularly when the risks are taken out of context of the risks associated with other fuel cycles.

COSTS OF ADDITIONAL CAPACITY IN DIFFERENT FUEL CYCLES

It is likely that Sweden will require investments in additional generation capacity over the coming two decades, and the required capacity expansions will be particularly significant if nuclear power is phased out. As was emphasized in the simple supply-and-demand model above, the cost of the nuclear phaseout will be primarily determined by the cost of replacement power along with the timing of the phaseout. In this section, we discuss the estimated costs of new power sources.

Assuming that new capacity can be brought on line in an orderly manner, with the necessary planning and efficient scheduling of new plants, then the cost of new capacity will be determined by three key elements: the choice of generation technique, the capital costs of new capacity, and the fuel price. The cost-driven elements are for the most part publicly available technologies or fuels, so the uncertainties about prices are relatively modest. For this study, we use data on the costs of electricity collected by the Nuclear Energy Agency, which comprise a careful survey of data from the major industrial countries.[22] This study gathered and analyzed data on the projected costs from nuclear, coal-fired, gas-fired, and renewable sources from utilities and government agencies in sixteen OECD and six non-OECD countries. The study has the advantage of using a standardized methodology, of allowing comparisons among the different countries, and of having a historical baseline so that we can compare different years.

The study focused on plants that could be available for commissioning in the year 2000. The nuclear technologies were primarily light-water reactors, whereas the coal-fired technologies were primarily pulverized fuel coal plants and the gas-fired plants were primarily combined-cycle gas turbines.

Tables 11 through 13 show the estimated costs of producing nuclear, coal-fired, and gas-fired power in different OECD countries, using a 5% real discount rate. For these figures, we have used market exchange rates rather than purchasing-power parity exchange rates because most of the technologies are tradable.[23]

To estimate the costs of generating electricity for Sweden required some adjustments of the survey. For the nuclear power costs, there are no estimates available for Sweden because it is currently unlawful even to plan for nuclear power expansion; for this technology, I have used estimates that apply to northern European countries. For coal-fired plants, I have used the estimates that were provided by Sweden in the survey; the coal plants costed for Sweden include extremely stringent pollution control techniques. These estimates seem reasonable and are in line with those of comparable European countries. Finally, for gas-fired plants, the Swedish response did not include an independent price for natural gas, so I have taken the prices that were estimated by northern European countries with no indigenous gas supplies. The estimated prices were quite consistent across these countries at $3.6 per gigajoule of gas (approximately $3.6 per million British thermal units of natural gas) and are in line with current delivered prices of gas.

Table 11. Levelized Electricity Generation Costs: Nuclear (cost in 1991 $U.S. per 1000 kWh).

Country	*Investment*	*Operations & maintenance*	*Fuel*	*Total*
Canada	22.8	5.3	1.8	29.8
Finland	18.9	5.4	5.8	30.1
France	14.5	10.0	8.3	32.8
Germany	29.6	12.7	10.8	53.1
Japan	24.4	10.9	18.3	53.7
U.K.[a]	31.5	10.5	8.1	50.0
US–1[b]	21.1	16.4	5.2	42.7
US–2[b]	22.1	16.4	5.2	43.7
US–3[b]	20.5	16.4	5.2	42.1

[a]Average of different estimates.

[b]U.S. is broken into three regions. US–1 is midwest, US–2 is northeast, and US–3 is west.

Notes on sources and methods underlying Tables 11–13: The underlying data source is OECD, Nuclear Energy Agency, *Projected Costs of Generating Electricity*, OECD, 1993. The study surveyed actual and projected costs in most OECD countries using a standardized methodology for different components of costs. Specific important assumptions are as follows:

1. The date of commercial commissioning is July 1, 2000.
2. The nuclear reactor type assumptions are:
   Canada: Pressurized-Heavy-Water Reactor (PHWR)
   Japan: Light-Water Reactor (LWR)
   Netherlands: Simplified Boiling-Water Reactor (SBWR)
   U.S.: Evolutionary Light-Water Reactor (ELWR)
   Others: Pressurized-Water Reactor (PWR)
3. The operating lifetime of the nuclear and coal-fired power stations are generally 30 years.
4. The settled-down load factor is 75%.
5. The discounted levelized load factor (for reference cases) is
   at a discount rate of 5% p.a.: 73.8%
   at a discount rate of 10% p.a.: 73.0%

Table 12. Levelized Electricity Generation Costs: Coal (cost in 1991 $U.S. per 1000 kWh).

Country	*Investment*	*Operations & maintenance*	*Fuel*	*Total*
Canada	14.40	4.10	15.50	34.00
Finland	9.80	6.70	18.60	35.00
France	11.70	9.50	29.40	50.60
Germany	16.90	15.10	48.10	80.10
Japan	20.60	7.90	34.50	63.00
U.K.[a]	17.75	12.10	19.40	49.20
US–1[b]	17.40	10.20	17.10	44.70
US–2[b]	17.70	9.60	24.00	51.30
US–3[b]	16.80	7.30	11.20	35.30

[a]Average of different estimates.

[b]U.S. is broken into three regions: US–1 is midwest, US–2 is northeast, and US–3 is west.

Source: See Table 11 for sources and methods.

Table 13. Levelized Electricity Generation Costs: Gas (cost in 1991 U.S. dollars per 1000 kWh).

Country	*Investment*	*Operations & maintenance*	*Fuel*	*Total*
Canada	8.2	2.2	41.8	52.2
Finland	5.8	4.0	25.5	35.3
France	7.0	4.2	43.6	54.8
Germany	na	na	na	na
Japan	12.7	6.9	57.7	77.3
U.K.[a]	6.5	6.5	32.3	45.2
US–1[b]	6.2	2.5	39.0	47.7
US–2[b]	6.2	2.5	42.4	51.1
US–3[b]	6.2	2.5	40.4	49.1

[a]Average of different estimates.

[b]U.S. is broken into three regions: US–1 is midwest, US–2 is northeast, and US–3 is west.

Source: See Table 11 for sources and methods.

Table 14 shows the estimated levelized costs of generating electricity that are derived from these sources and that I will apply to Sweden in the estimates below. The estimated costs at a 5% discount rate ranged from 4.4 cents per kWh for electricity generated with natural gas to 5.4 cents per kWh for coal-fired plants. At a discount rate of 10%, the cost of generation ranged from a low of 5.1 cents per kWh for natural gas to a high of 7.0 cents per kWh for both nuclear and coal generation.

Outside of nuclear power, the major uncertainties for electricity cost in the near term concern fuel prices. Fuel prices comprise approximately one-third of the cost for coal and more than one-half of the cost for gas. Coal prices have been relatively stable and are competitively priced in international markets. Natural gas, by contrast, has significant monopoly elements because of the high transportation costs. For Sweden, the current supplier is Denmark, although Norway and Russia are the major potential future suppli-

Table 14. Assumptions for Cost of Replacement Power in Sweden (all prices in 1995 $U.S.).

	Capital costs (per kWe) at discount rate		*Levelized costs (dollars per 1000 kWh)*					
			Capital costs at discount rate		*O&M*	*Fuel*	*Total costs at discount rate*	
	5%	*10%*	*5%*	*10%*			*5%*	*10%*
Coal	2098	2334	20.6	37.0	9.7	23.7	54.0	70.4
Gas	955	1063	9.4	16.9	6.1	28.1	43.6	51.1
Nuclear	2769	3155	27.2	50.0	10.6	9.9	47.7	70.5
Average:								
Three technologies							48.4	64.0
Two fossil technologies							48.8	60.8

Source: Nuclear Energy Agency, *Projected Costs of Generating Electricity, Update 1992*, OECD, Paris, 1993. Estimates for coal and gas are for Sweden with the exception of fuel costs for gas. For gas costs, the estimates were the average gas price for Belgium, Denmark, Finland, and France (which was $3.6 per GJ of gas) using a thermal efficiency of 51 percent. For nuclear power, estimates were the weighted average of estimated costs for Germany, Belgium, and Finland, with the weight on Germany being twice that of the other countries. Underlying data are given in this reference in Tables 9, 10, 19, 21, and 22.

ers. Concerning nuclear power, major uncertainties remain for most countries, but this is not an issue in Sweden given the slim possibility of building new nuclear power reactors in the next few years.

While the estimates for fossil fuel costs are probably relatively reliable, the nuclear power costs are subject to serious doubt. The survey shown in Table 14 is generally based on *projections* of nuclear costs rather than on recent experience. One of the discouraging facts is that actual costs have overrun projections, and that costs have been rising almost continually since the first nuclear power plants were built. Figure 1 shows the "overnight costs" of nuclear plants in different countries. This figure indicates that outside of France, costs escalated sharply, particularly in the U.S. and U.K.; moreover, the costs may well be above the estimates shown in Table 14.[24] This reservation does not affect the basic results of this study, however, because the replacement power estimates used here do not involve nuclear power.

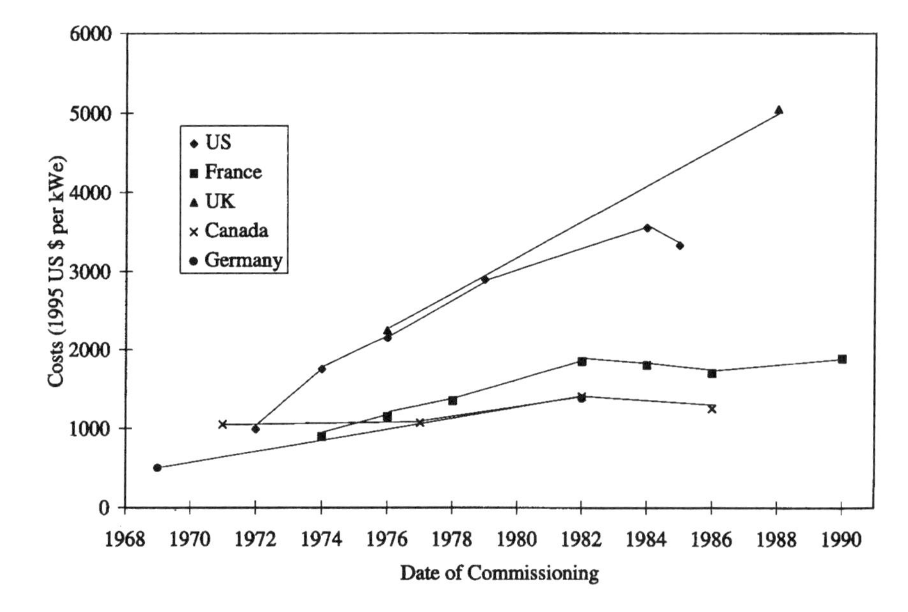

Figure 1. Capital Costs of Nuclear Power.

Note: Costs are estimated or actual "overnight" costs, not the projected costs that are used in Table 14.

ENDNOTES

[1]See Lennart Hjalmarsson, "From Club Regulation to Market Competition," in Richard Gilbert and Edward Kahn, eds., *International Comparisons of Electricity Regulation*, New York, Cambridge University Press, 1996.

[2]The use of nominal interest rates rather than real interest rates tends to tilt the cost figures so that the capital costs are relatively high in the early part of a plant's accounting lifetime and relatively low in the latter part of a plant's accounting lifetime. For this reason, the total costs are likely to be understated.

[3]Data in this section are drawn from NUTEK, *Energy in Sweden: Facts and Figures, 1995.*

[4]Nuclear Energy Agency, *Nuclear Power Economics and Technology: An Overview*, OECD, 1992, p. 78f.

[5]Forsmark, *Annual Report: 1994.*

[6]The discussion in this section is based on Nuclear Energy Agency, *Decommissioning of Nuclear Facilities*, Paris, 1991.

[7]As in other sections, all values are converted into 1995 U.S. dollars using the chain-weighted gross domestic product deflator.

[8]Ibid., Nuclear Energy Agency 1991, p. 51.

[9]Ibid., Table 11.

[10]Discussion of SKB's activities are contained in SKB, *Activities, 1993*, Stockholm.

[11]There is mixed evidence for the United States on the time trends in availability of nuclear power as the plants age. Komanoff found that availability rates rise with age and interprets this as a learning effect (Charles Komanoff, *Power Plant Performance Update 2*, Komanoff Energy Associates, New York, 1978). A more recent study by Krautmann and Solow found no significant learning after six years and some indication for BWRs that availability declines with age after six years (see Anthony C. Krautmann and John L. Solow, "Nuclear Power Plant Performance: The Post Three Mile Island Era," *Energy Economics*, July 1992, pp. 209–216).

[12]U.S. Nuclear Regulatory Commission, *Reactor Safety Study: An Assessment of Accident Risks in U.S. Commercial Nuclear Power Plants*, WASH-1400, NUREG-75/014, October 1975, hereafter *Reactor Safety Study.*

[13]See particularly *Report of the President's Commission on The Accident at Three Mile Island*, Pergamon Press, 1979.

[14]The estimates in the paragraph are cited in Nuclear Energy Agency, *Broad Economic Impact of Nuclear Power*, OECD, Paris, 1994.

[15]The discussion in the paragraph is drawn from Lars Högberg, "The Swedish Regulatory Approach to Nuclear Safety," Presentation at the Symposium on Nuclear Power, Bangkok, Thailand, October 1993.

[16]To calculate the maximum likelihood from the historical data, we use the probability-severity relationship from *Reactor Safety Study*. The maximum likelihood estimate, assuming that the frequency of severe accidents follows a Poisson distribution, is the inverse of the number of reactor years.

[17]David Fischer, "Nuclear Non-proliferation," *Energy Policy*, July 1992, pp. 672–682.

[18]Ibid., p. 681.

[19]Theodore B. Taylor, "A Ban on Nuclear Technologies," *Technology Review*, August/September 1995, pp. 76–77.

[20]See U.S. Environmental Protection Agency, *The Benefits and Costs of the Clean Air Act, 1970 to 1990*, May 3, 1996, Draft, Washington, D.C. This landmark study is the first comprehensive valuation of the economic impact of environmental regulation. The study uses a value of $4.8 million per life saved in 1990 prices. This figure is adjusted in the present study to reflect alternative estimates and certain methodological issues in the EPA study. In any case, because the final estimate of the external costs of electricity are so small, the exact figure for the value of life has little impact on the estimates of the cost of a nuclear phaseout.

[21]An alternative estimate of the external costs of electricity generation for Germany is contained in Rasiner Friedrich and Alfred Voss, "External costs of electricity generation," *Energy Policy*, February 1993, pp. 114–122. Their estimates of external costs (converted at 1.5 DM per U.S. dollar in 1988 prices) are 0.44 to 1.68 cents per kWh for coal and 0.03 to 0.17 cents per kWh for nuclear power. The estimates provided above are comfortably inside the Friedrich-Voss estimates.

Many studies of the external costs of different power systems include occupational accidents and fatalities in mining and other parts of the fuel cycles. This inclusion is questionable if workers are aware of the potential riskiness of these occupations and work in them in return for a differential pay that compensates them for the increased risk.

[22]Nuclear Energy Agency, *Projected Costs of Generating Electricity: Update 1992*, OECD, Paris, 1993.

[23]The relative merits of PPP versus market exchange rates were discussed in Chapter 2.

[24]These data are drawn primarily from Gordon MacKerron, "Nuclear Costs: Why Do They Keep Rising?" *Energy Policy*, July 1992, pp. 641–652. "Overnight costs" refers to the total cost of building a plant as if all costs were incurred at one moment in time. It therefore removes inflation and interest costs on construction.

7

Alternative Scenarios for Nuclear Power

This chapter puts the different pieces of the puzzle together to look at the total picture regarding the Swedish nuclear phaseout. Here, I present my estimates of the costs of different scenarios for Swedish nuclear power using the Swedish Energy and Environmental Policy (SEEP) model. We take as a baseline case for reference purposes a scenario in which the current plants continue to operate for forty-year lifetimes. Assuming a forty-year lifetime, the weighted average date of decommissioning for Sweden's twelve nuclear power plants would be 2020. We do not allow for the possibility of new nuclear construction, but under the baseline assumptions it appears further nuclear power stations would probably not be built because other technologies, particularly gas-fired plants, appear to be more economical. The major foreseeable difference among the technologies would be that, in the event of very stringent limitations on carbon dioxide emissions, the fossil technologies would become much more expensive and nuclear power would become economically advantageous.

In this chapter, we examine a number of different possible nuclear policies. The major policy of interest is the *nuclear phaseout*. For this case, all nuclear power plants are phased out in 2010. This should be taken to represent a case in which the *average* shutdown date is 2010, rather than a literal closing down of all twelve plants on the same date; in other words, this policy represents a shutdown ten years before the estimated economic lifetime. In addition, in this chapter I analyze both the interaction of the nuclear power decisions with Sweden's commitments on climate change as well as other options and perform a sensitivity analysis on the approach used here.

The method by which I estimate the economic impacts of alternatives is as follows: We begin with the baseline projection of the SEEP model described above. Recall that the baseline allows current nuclear power plants to operate to the end of their economic lifetimes; it therefore represents the economically efficient policy in the absence of overriding environmental policies. We then modify the policy constraints or estimated costs of different technologies to conform to different assumptions and run the SEEP model with the modified assumptions. Finally, we compare the economic results for the baseline with the modified path. The difference in the present value of income between the two paths is the net economic impact of the policy.

SOURCES OF REPLACEMENT POWER

It will be useful to step back and discuss the sources of replacement power if Sweden does indeed phase out its nuclear power. There are four ways to balance supply and demand after a nuclear phaseout:

- raise prices sufficiently to fit consumption into production,
- produce electricity from domestic fuels,
- import electricity, or
- produce electricity from imported fuels.

The first two may be ruled out as unrealistic. For the first, the price increase necessary to balance demand and nonnuclear supply would cause havoc in Sweden and is surely politically unacceptable. For the second option, there are insufficient domestic fuel supplies except for biofuels, which are expensive and pose serious environmental problems at the required scale.

The third and fourth options (or some combination of them) raise more interesting issues. In terms of importing electricity, the only source of supply that might compete with domestic production would be Norway, using either Norwegian hydroelectric power or electricity generated with Norwegian natural gas. Rough orders of magnitude suggest that these are technically feasible. Norwegian hydroelectric power is capable of a 40% expansion, which would constitute about 50 terawatt-hours (tWh) compared to the 70 tWh generated by nuclear power in Sweden. In addition, Norway has substantial supplies of natural gas. There have been some rumblings in the Scandinavian press that Norwegian gas reserves are fully committed, but this may be a prelude to tough bargaining on the part of the Norwegians.

In terms of importing fuels, petroleum is an obvious but high-cost and high-risk source. Coal is lower cost and has less political risk, but it has environmental costs that might be unacceptable for Sweden (see particularly the risks shown in Tables 9 and 10 in Chapter 6). Natural gas is the obvious choice for an imported fuel from an environmental and economic point of view. At present no major natural gas pipelines extend to Sweden. Two potential sources of supply are Norway (discussed in the last paragraph) and Russia over the longer term. Russia clearly has adequate natural gas supplies, but the distances, costs, and political risks are daunting.

In the longer run, it would seem unwise for Sweden to rely too heavily on a single source of replacement power, whether that be oil, natural gas, or imported electricity. The risks of monopoly prices, an adverse bargaining position, and political risks (particularly in Russia) would seem too great to hold the country hostage to a single supplier. A prudent policy would divide the replacement power among three or four sources. A possible portfolio of sources might be to aim for new electric power to be drawn in roughly equal measure from Norwegian hydroelectric power; Norwegian gas, coal, or oil; and Russian gas. The difficulty of ensuring a balanced source of supply in a deregulated electricity market poses yet another issue that would need to be dealt with. These considerations suggest that replacing nuclear power will be no easy matter and that Sweden should count on paying the world price of replacement fuels—that is, the British thermal unit–equivalent price of oil for fuels or the replacement cost of electricity from imported electricity.

MAJOR OPTIONS

Total Phaseout in 2010 ("2010 Phaseout")

The first policy is a case that corresponds to the date envisioned in the referendum and imposes a total phaseout of nuclear power in the year 2010. Because we assume that the plants will have a remaining lifetime of ten years in 2010, this policy is equivalent to shutting down the nuclear plants on average ten years before the end of their useful economic lifetimes. It is recognized that all plants would not be shut on exactly the same date, but a uniform date is imposed for computational simplicity. Since the model assumes perfect foresight and a smoothly adjusting world, this would be virtually equivalent to a phaseout that was *centered* on 2010.

Because it calculates the long-run equilibrium,[1] the SEEP model assumes that businesses and power producers fully anticipate the timing and consequences of the phaseout. Given realistic frictions, the economic impact is likely to be underestimated. For example, if the discontinuation takes place before new baseload plants can be economically planned, constructed, and put in operation, then the costs of dislocation are likely to be higher than those presented here. Or, if the price consequences are misestimated, then investments may be inefficiently designed. The economic costs of such mistakes are difficult to assess and are not included in the estimates.

A summary comparison of the baseline run with the Phaseout 2010 run is shown in Table 1. This table compares the important variables in the year 2010 for the baseline run with the case in which nuclear power is phased out in 2010. Electricity prices will rise and electricity use will decline sharply after the phaseout as only 60 of the 70 tWh of generation due to the nuclear phaseout will be made up from other sources. The SEEP model foresees the production being replaced with gas-fired plants, but it could easily be replaced in part with oil-fired and coal-fired plants if environmental laws permitted or with imported electricity if that were available.

The SEEP model predicts that a nuclear phaseout in 2010, under conditions where it is anticipated and efficiently adapted to, will lead to substantial economic losses. The discounted value of the total loss from 1994 to 2030 (discounted to 1995 and in 1995 prices) will be about $8.9 billion. Annualizing this at a 5% discount rate yields an annual loss of $0.43 billion per year. These calculations estimate the cost of the phaseout by discounting back to 1995 rather than at the date of the phaseout. This convention is for convenience and consistency, but a better measure would be to measure the impact as of the date of the phaseout, which in this case is 2010. At that date, and using a 5% discount rate, the cost is higher by a factor

Table 1. Summary Results for Baseline Run and Nuclear Phaseout, 2010.

	Baseline	*Nuclear phaseout in 2010 with no CO_2 limitation*
Energy demand (tWh)	706.4	687.6
Specific electric	85.0	80.7
Transportation	136.9	136.7
Other	484.5	470.2
Electricity production (tWh)		
Sources	142.0	132.1
Nuclear	70.2	0.0
Hydroelectric	67.8	67.8
Other existing	4.0	6.2
New capacity	0.0	58.1
Environmental indexes		
CO_2 emissions (millions of tons CO_2)	76.1	98.7
SO_2 emissions (thousands of tons SO_2)	112.7	113.6
Energy prices (U.S. cents per kWh, 1995 prices)		
Electricity		
Bus bar	4.39	4.83
Industrial	6.77	7.20
Residential	12.37	12.81
Transportation	11.84	11.84
Other	4.04	4.08
Externality prices ($U.S.,1995 prices)		
CO_2 (per ton CO_2)	None	None
SO_2 (per ton SO_2)	None	None
Externality limits		
CO_2 (millions of tons CO_2)	None	None
SO_2 (thousands of tons SO_2)	None	None
Energy-Output Ratio (1994 = 100)		
Electricity-GDP	75.53	70.27
Energy-GDP	90.95	88.53

Note: Table shows the baseline run for the SEEP model along with the results of a nuclear phaseout in 2010. Other detailed results for the baseline run are shown in Table 1 in Chapter 5.

of $1.05^{15} = 2.08$, or approximately double. Hence, the cost of the phaseout at the date of phaseout (that is, in 2010) is $18.4 billion.

There is, in this phaseout scenario, a substantial increase in emissions of carbon dioxide (CO_2) but, because the replacement fuel is natural gas, little change in the emissions of sulfur dioxide (SO_2). It is estimated that SO_2 rises 1% as the electricity system moves from a nuclear-based to a fossil system. CO_2 emissions rise sharply, by 30% compared with the baseline case in 2010. With respect to international commitments, CO_2 emissions of ninety-nine million tons with a nuclear phaseout are almost double the 1990 levels, whereas Sweden has set a national target of maintaining CO_2 emissions at 1990 levels or below with a decline in emissions after 2000.

A crucial assumption in this scenario is that electricity generation will have time to adapt to the nuclear phaseout. The costs are likely to be significantly larger than those estimated here if the nuclear discontinuation is applied without adequate lead time for replacement power to be found.

Limiting CO_2 Emissions to 1990 Levels

We discussed above the growing concerns about climate change. At the Rio Summit of 1992, nations agreed to limit emissions to 1990 levels by the year 2000, and the Berlin Conference of the Parties in 1995 agreed to extend the limitation beyond the year 2000. Sweden has a national objective of limiting CO_2 emissions to 1990 levels, although it is recognized that this target may be difficult to meet and might be inefficient compared with finding low-cost emissions reductions in other regions of the world.

Because of the importance of the climate-change commitment, we investigated two scenarios in which Sweden meets its national objectives for CO_2 emissions reductions. In the first scenario (CO_2 limitation), CO_2 emissions were limited to 1990 emissions rates for all future years, but there was no phaseout of nuclear power. In the second scenario (combination nuclear phaseout and CO_2 limitation), CO_2 emissions were limited as in the first scenario and in addition nuclear power was shut down in the year 2010.

Table 2 compares the baseline scenario with the CO_2 limitation scenario while Table 3 shows the results of the combined nuclear

Table 2. Summary Results for Baseline Run and CO_2 Limitation, 2010.

	Baseline	*Limit CO_2 emissions to 1990 quantities*
Energy demand (tWh)	706.4	634.8
Specific electric	85.0	79.2
Transportation	136.9	107.9
Other	484.5	447.7
Electricity production (tWh)		
Sources	142.0	138.0
Nuclear	70.2	70.2
Hydroelectric	67.8	67.8
Other existing	4.0	0.0
New capacity	0.0	0.0
Environmental indexes		
CO_2 emissions (millions of tons CO_2)	76.1	50.5
SO_2 emissions (thousands of tons SO_2)	112.7	76.5
Energy prices (U.S. cents per kWh, 1995 prices)		
Electricity		
Busbar	4.39	4.78
Industrial	6.77	7.15
Residential	12.37	12.76
Transportation	11.84	14.64
Other	4.04	4.71
Externality prices ($U.S., 1995 prices)		
CO_2 (per ton CO_2)	None	112
SO_2 (per ton SO_2)	None	None
Externality limits		
CO_2 (millions of tons CO_2)	None	50.5
SO_2 (thousands of tons SO_2)	None	None
Energy-Output Ratio (1994 = 100)		
Electricity-GDP	75.53	73.40
Energy-GDP	90.95	81.73

Note: Table compares the baseline run with the CO_2 limitation run. The major difference is that CO_2 emissions in Sweden are limited to the 1990 level.

Table 3. Summary Results for Baseline Run and Combination Nuclear Phaseout and CO_2 Limitation, 2010.

	Baseline	*Nuclear phaseout and limit CO_2 emissions to 1990 quantities*
Energy demand (tWh)	706.4	509.0
Specific electric	85.0	49.7
Transportation	136.9	101.4
Other	484.5	357.9
Electricity production (tWh)		
Sources	142.0	74.3
Nuclear	70.2	0.0
Hydroelectric	67.8	67.8
Other existing	4.0	0.0
New capacity	0.0	6.5
Environmental indexes		
CO_2 emissions (millions of tons CO_2)	76.1	50.5
SO_2 emissions (thousands of tons SO_2)	112.7	72.9
Energy prices (U.S. cents per kWh, 1995 prices)		
Electricity		
Bus bar	4.39	9.73
Industrial	6.77	12.11
Residential	12.37	17.71
Transportation	11.84	15.23
Other	4.04	5.24
Externality prices ($U.S., 1995 prices)		
CO_2 (per ton CO_2)	None	135
SO_2 (per ton SO_2)	None	None
Externality limits		
CO_2 (millions of tons CO_2)	None	50.5
SO_2 (thousands of tons SO_2)	None	None
Energy-Output Ratio (1994 = 100)		
Electricity-GDP	75.53	39.52
Energy-GDP	90.95	65.54

Note: Table compares the baseline run with one that combines a nuclear phaseout with a CO_2 emissions limitation.

phaseout and CO_2 limitation. These runs show that limiting CO_2 emissions will have a major impact on the cost of the nuclear phaseout. Looking at the most stringent scenario in Table 3, there is a significant curtailment of the energy and electricity sectors in the year 2010. Electricity production is reduced almost one-half relative to the baseline scenario, and total energy production is reduced almost one-third. Industrial electricity prices rise more than 50% above the baseline level, and the higher level (12.1 cents per kilowatt-hour [kWh]) electricity prices are almost triple their levels in 1990.

The present value of the impact for the combined CO_2 limitation and nuclear phaseout is a reduction of real income of $53 billion relative to the base case (again in 1995 prices and discounted to 1995). If we annualize the impact of the most stringent case, the impact on economic welfare is substantial, with a reduction of $2.6 billion, or a little more than 1% of current gross domestic product (GDP). Note that impact will be a larger fraction of income at the time of the phaseout because the cost (which is carried forward at 5% per year) will grow faster than income (projected to grow at 2% per year).

Table 4 provides a summary of the economic impacts of alternative policies for the four major cases analyzed in this study. The four cases are the baseline (shown in the northwest quadrant of the table), the nuclear phaseout (shown in the southwest part), the CO_2 limitation case (shown in the northeast), and the combined nuclear phaseout and CO_2 limitation (shown in the southeast).

The results indicate that limiting CO_2 emissions has a social cost to Sweden approximately five times as high as the nuclear shutdown in 2010, being $35 billion for the CO_2 limitation compared to $9 billion for the nuclear phaseout. However, imposing *both* of the environmental policies leads to a cost that is greater than the sum of the parts, equal to $53 billion in present value for the combined policies. In other words, the cost of a nuclear phaseout is approximately twice as high in the presence of the CO_2 constraint.

Gradual or Straight-Line Phaseout from 2000 to 2010

We next turn to two alternative phaseout policies. One alternative, which phases out nuclear power somewhat more quickly, is a "straight-line" phaseout between 2000 and 2010. This policy

Table 4. Estimates of Economic Impact of Nuclear Phaseout: Shutdown of All Plants in 2010 (1995 $U.S., billions, discounted to 1995).

	No CO_2 limitation	*CO_2 limitation at 1990 emissions*
A. Discounted value of objective function at 5 percent real discount rate		
With nuclear power	5,625	5,589
No nuclear power after 2010	5,616	5,572
B. Difference from baseline run (difference from upper left entry)		
With nuclear power	0.0	–35.4
No nuclear power after 2010	–8.9	–52.6

Note: The estimates of the cost of a nuclear phaseout were calculated in the SEEP model under four sets of assumptions: Base (northwest), nuclear phaseout (southwest), CO_2 limitations (northeast), and nuclear phaseout with CO_2 limitations (southeast). The precise definitions of the different runs are: *Base*, nuclear power is allowed in all years throughout their economic lifetime of 2020; *Nuclear phaseout*, all nuclear power plants are shut down in 2010; *CO_2 limitation*, nuclear power is allowed to operate until its economic lifetime of 2020, but CO_2 emissions are limited to 1990 emissions rates; *Nuclear phaseout with CO_2 limitations*, combination of CO_2 limitation and nuclear phaseout.

assumes that the plants are shut down gradually over the period. More precisely, the straight-line case assumes that capacity is reduced by 10% of initial capacity in each year. Hence, capacity is reduced by 7 tWh in 2001, 14 tWh in 2002, and so forth until a total 70-tWh reduction in 2010.

The effect of this accelerated phaseout is, not surprisingly, a higher cost than the 2010 phaseout. The total cost rises from $8.9 billion to $15.0 billion, for a 70% increase in total cost. The reason the cost is so much higher is that the useful lifetime has been shortened by fifteen years (instead of ten years in the 2010 phaseout case), while the extra five years of shortening occurs closer to the present, which increases the discounted value more than proportionally.

Phase Out Only Two Oldest Plants in 2000

One possible strategy is to limit the phaseout to two of the oldest plants and then to allow the rest to operate for their useful lifetime.

The rationale behind this strategy might be that it is a compromise that would satisfy the spirit of the nuclear referendum by shutting down the oldest (and presumably the least safe) of the plants, while, as current Swedish policy emphasizes, not excessively harming Sweden's standard of living or international competitiveness.

To implement this run, I selected the two oldest reactors with a total rated capacity of 1,045 megawatts. Removing these reactors reduced the total nuclear generation over the 2000–2020 period by about 10%—from 70.2 tWh per year to 63.1 tWh per year. The estimated impact of this shutdown is $0.83 billion (again, discounted to 1995). This is clearly significantly less than the cost of the entire shutdown. The reasons for the lower costs are three: first, these two reactors represent only 10% of the rated capacity; second, their useful lives would be less than those of the other reactors; and, third, the last two reactors are the marginal source of power and are therefore less valuable to the nation than the average power source.

Figure 1 shows the estimated costs of a phaseout for the four nuclear policies that are examined here. In each case, we measure the cost relative to the no-phaseout or baseline case. The costs range from a low of $0.8 billion for the policy phasing out the two

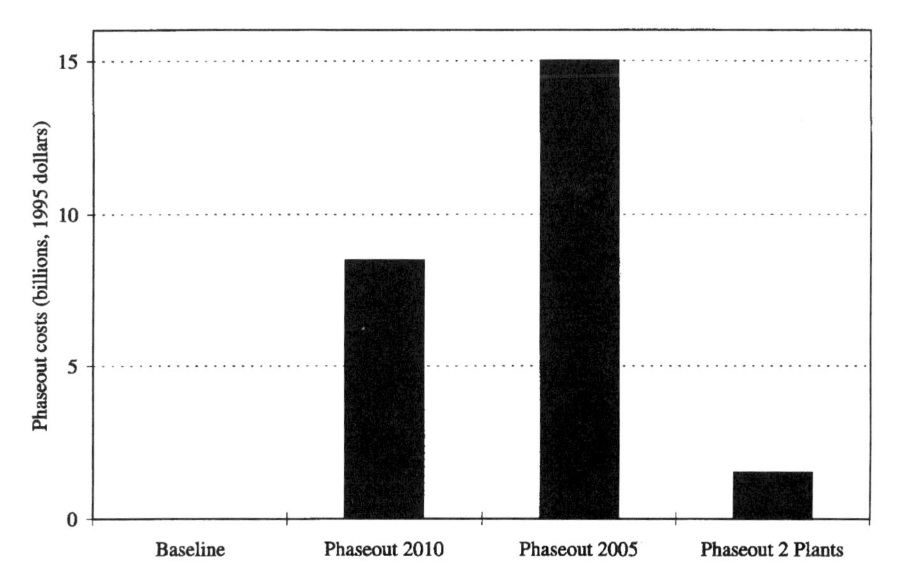

Figure 1. Phaseout Costs of Alternative Policies.

oldest reactors in 2000 to about $15 billion for the straight-line phaseout between 2000 and 2010.

SENSITIVITY ANALYSES

Models like that employed here are generally sensitive to the assumptions that go into them. I have therefore undertaken an extensive sensitivity analysis to determine the impact of alternative assumptions about the various important elements of the energy model and of energy and environmental policy. These include alternative assumptions about the costs of new technologies, alternative assumptions about the discount rate, more optimistic assumptions about conservation and renewable energy resources, alternative assumptions about the growth of output, and an augmented model that allows greater international trade in electricity. The results are collated in Tables 5 and 6.

Alternative Cost Assumptions

One of the major uncertainties about the future is the movement of oil and gas prices. It is common wisdom today that oil and gas prices are unlikely to rise sharply in the next decade or so given the ample supplies and prospects of large new natural gas supplies coming on line. An alternative view is that because of unforeseeable economic or political events, oil and gas prices may rise sharply in the coming years. This is obviously an important question because the replacement source of electricity is likely to be predominantly fossil fuels. To test for the impact of this contingency, I estimated the impact of a doubling of oil and gas prices; at the same time, coal and biofuel prices were assumed to rise by 25% because of the induced demand increase. These did indeed lead to an increase in the cost of a nuclear phaseout in 2010, with the cost rising from $8.9 billion to $13.0 billion.

An additional possibility is that the cost of constructing new facilities will prove higher than the estimates presented here. Sweden has not constructed large-scale energy facilities for over a decade, and many countries have found their capital costs rising sharply during that period, largely for environmental reasons.[2] We therefore tested the impact of doubling the capital costs of new

Table 5. Value of Objective Function for Alternative Runs (billions of 1995 $U.S. discounted to 1995).

	No phaseout	*Complete phaseout in 2010*	*Phase out from 2000 to 2010*	*Phase two reactors out in 2000 only*
Base parameters				
No CO_2 policy	5624.7	5615.8	5609.7	5623.9
With CO_2 policy				
Limit to 1990 emissions	5589.3	5572.0		
Additional 20 percent reduction	5567.9	5550.2		
1990 limitation plus offsets	5614.4	5603.3		
Sensitivity analyses				
Deregulation				
International price at 5.3 cents per kWh	5625.2	5616.3		
International price at 4.4 cents per kWh	5625.2	5616.6		
Alternative costs				
Capital recovery rate doubled	5608.7	5597.6		
Oil and gas prices doubled	5467.0	5454.0		
Discount rate of 10%	3154.3	3150.6		
Conservation and renewables				
30% conservation at 2.6 cents per kWh	5630.9	5623.2		
Renewables at 1 tWh/yr/yr	5625.2	5617.6		
Parameters				
Economic growth +0.5%	6084.9	6076.0		
Economic growth –0.5%	5219.3	5210.5		
Lifetime at 50 years	5644.6	5630.2		
20 tWh of new hydroelectric	5625.2	5617.6		

Note: Table shows the present value of the objective function in different runs. Note that runs are not always comparable and that comparisons should be made horizontally between case with and without a nuclear phaseout.

Table 6. Impact of a Nuclear Phaseout for Alternative Runs (billions of 1995 $U.S. discounted to 1995).

Sensitivity Run		*No phaseout*	*Complete phaseout in 2010*	*Phase out from 2000 to 2010*	*Phase two reactors out in 2000 only*
	Base parameters				
B	*No CO_2 policy*	0.0	–8.9	–15.0	–0.8
	With CO_2 policy				
1	Limit to 1990 emissions	0.0	–17.2		
2	Additional 20 percent reduction	0.0	–17.7		
3	1990 limitation plus offsets	0.0	–11.1		
	Sensitivity analyses				
	Deregulation				
4	International price at 5.3 cents per kWh	0.0	–8.9		
5	International price at 4.4 cents per kWh	0.0	–8.7		
	Alternative costs				
6	Capital recovery rate doubled	0.0	–11.1		
7	Oil and gas prices doubled	0.0	–13.0		
8	Discount rate of 10%	0.0	–3.7		
	Conservation and renewables				
9	30% conservation at 2.6 cents per kWh	0.0	–7.7		
10	Renewables at 1 tWh/yr/yr	0.0	–7.6		
	Parameters				
11	Economic growth +0.5%	0.0	–8.9		
12	Economic growth –0.5%	0.0	–8.8		
13	*Lifetime at 50 years*	0.0	–14.4		
14	*20 tWh of new hydroelectric*	0.0	–7.6		

Note: Table shows the difference in present value of the objective function in different runs. These are calculated as the difference of objective function between the phaseout and no-phaseout runs in Table 5. Values are in 1995 prices and are discounted back to 1995.

electricity generation sources as well as a 50% increase in the decommissioning costs. These turned out to have similar effects to the fuel price sensitivity analysis. The cost of a 2010 phaseout was in this case estimated to be $11 billion, or about one-fourth higher than the baseline case.

Higher Cost of Capital and Discount Rate

Another issue concerns the appropriate discount rate and cost of capital to use for estimating the impact of a nuclear phaseout. Historically, rates of return of 5% per year (inflation corrected) have often been used for cost-benefit studies of energy systems. In the current environment, however, Sweden is facing severe capital constraints because of its poor credit position in international markets and its high and growing public debt. In one sense, the Swedish economic crisis spills over into energy policy in a higher cost of capital for both private and, especially, public funds. To test for the sensitivity of the estimates to the capital cost, I examined the impact of a higher cost of capital—10% per year instead of the base 5% per year (see Table 14 in Chapter 6 for the estimated costs of new generation for different discount rates). This higher rate was used to increase the capital recovery factors for new plants as well as to discount future costs and benefits.

The estimated impact of a higher discount rate was, not surprisingly, to lower the discounted costs of a nuclear phaseout. At the higher discount rate, the cost was $3.7 billion rather than the $8.9 billion in the baseline comparison (discounted to 1995). This is quite misleading, however, because the higher discount rate is masking the impact on real living standards. A better way to calculate the impact is to ask about the annualized loss in economic welfare at the date of the phaseout.[3] The baseline case represents an annual loss of $0.93 billion per year for a 2010 phaseout. By contrast, the estimated loss with a 10% cost of capital is $1.5 billion per year from 2010 on. This represents an increase of two-thirds in the real annualized cost of a phaseout from the higher discount rate.

Optimistic Assumptions about Conservation and Renewables

The baseline case assumes that there will be no major breakthrough in either renewable electric technologies or energy conservation.

Many proponents of the nuclear phaseout look to alternative energy sources and conservation as the route to both a more secure energy system and an inexpensive nuclear phaseout.

What are the technical possibilities? One is a massive increase in the production of electricity from unconventional renewable sources—wind, biofuels, geothermal, and wave energy being possibilities often mentioned for Sweden. For this experiment, I assumed that Sweden could increase its renewable (nonhydroelectric) production by 1 tWh per year starting in 1994 to a total of 37 tWh by 2030. The assumed cost is 4.3 cents per kWh in 1995 prices. This figure is extremely optimistic as a projection of the costs of electricity generation in the absence of subsidies but is used in the spirit of choosing a low-cost sensitivity case. The impact of this assumption on the cost of the nuclear phaseout was positive but relatively small: the estimated cost of a phaseout was 14% less than the baseline cost.

An alternative question involves the possibility for significant increases in energy conservation—this has been a high priority in Sweden as in many other countries. To test for the sensitivity of the assumptions concerning conservation, I assumed that a major increase in low-cost energy efficiency improvements in the electricity-using sectors became available. This was accomplished by increasing electricity efficiency by 30% over the next twenty years at a cost of 2.6 cents per kWh replaced. The cost of conservation is drawn from a wide variety of studies concerning the difference between best-practice and current techniques. This can be rationalized by assuming that there is a 30% shortfall of actual behind best-practice techniques and that this shortfall will be reduced to zero at low cost over the next twenty years.

The impact of this assumption on the cost of the nuclear phaseout was not negligible. The overall cost of the phaseout in 2010 was reduced by 13%, from $8.9 to $7.7 billion. The assumption lying behind this case is problematical, however. If such investments are worthwhile, they would be almost equally valuable *without* a nuclear phaseout as with it. It is difficult to see why Sweden would be spurred to make a massive effort in conservation without nuclear power that it would not make with nuclear power intact. If the conservation potential were exploited in either case, then the reduction in phaseout cost disappears.

Alternative Growth Projections

Another set of assumptions that might be crucial to the analysis involves economic growth. The estimates used here rely on medium-term projections made by the Swedish government. They are higher than those used by NUTEK and by some environmental groups, although they are lower than those prepared as background for this study. The estimated economic growth probably has an uncertainty range of plus or minus 0.5% per year over the next two decades, and we tested the impact of this range on the cost of a nuclear phaseout. The impact was surprisingly small, with the cost of a phaseout being 0.005% larger in the case of the higher economic growth and 2% smaller in the case of the lower economic growth.

The reason why economic growth has so little effect on the cost of a nuclear phaseout was shown in the simple supply-and-demand example in an earlier section: Once the price has reached the cost of replacement electricity, the cost of a nuclear phaseout is determined primarily by the cost of replacement power, and the growth of demand for power therefore has little effect on the cost of a phaseout.

Trade in Electricity in a Deregulated European Market

One of the major uncertainties facing the Swedish electricity market in the coming years is the pace of deregulation and liberalization, both internally and in the wider European market. Earlier sections discussed the major issues; in those sections, we suggested that to a first order of approximation there would be little interaction between deregulation and the costs of a nuclear phaseout. This section provides some quantitative estimates of the interaction of deregulation with the nuclear phaseout question.

In the context of an integrated European market it is not unrealistic to believe that imports of electricity could satisfy a substantial fraction of the demand displaced by curtailed Swedish nuclear power generation. The realistic difficulty arises, however, that there would be great incentives for any low-cost producer, such as Norway, either to price at a level close to that of Swedish replacement power or to charge significant monopoly rents on its exports.

Through the pricing of electricity exports, the Norwegian decision-makers are likely to raise the price to a level close to that of the next best alternative source of supply, thereby reducing the economic attractiveness of the Norwegian import option. To be sure, it seems likely that barriers to international trade in electricity will decline in the coming years, and this may provide opportunities for replacing some of the displaced demand if nuclear power is phased out. There seems little prospect, however, that a substantial fraction of the nuclear generation can be produced at prices that are significantly below the cost of alternative sources of domestic supply.

A rigorous approach to analyzing deregulation is extremely difficult as it requires a full model for the electricity sectors of all countries, for all transmission linkages, for each country's industrial organization, and for the foreign trade and pricing strategies that countries follow. Moreover, such an analysis must include all countries that are directly or indirectly linked together in the electricity market. Attempts to design such a model are under way. For our purposes, we can provide a rough approximation of the impact of a deregulated environment by assuming an international market in electricity that is perfectly competitive and frictionless. Because reality lies somewhere between the autarkic approach taken in the base case (or, strictly speaking, the case in which trade is independent of prices) and the frictionless free-trade model analyzed here, it is likely that the impact of a realistic trade environment lies somewhere between these two extremes.

For the free-trade case, I assume that there is a perfectly competitive international spot market for electricity in which Sweden is a price taker. I further assume that the transmission costs are zero, so that the Swedish price is equated to the international price (within the limits of capacity of international transmission links). Finally, I assume a range of international electricity prices that bracket the estimated cost of new capacity. In an earlier section, I estimated the marginal cost of new capacity to be about 5 cents per kWh, so for this experiment I take a high international price of 5.3 cents per kWh and a low price of 4.4 cents per kWh to estimate the impacts. In each case, I assume that exports are limited to 5 tWh while imports are limited to 20 tWh, and that the deregulation begins in 1998. These seem realistic parameters given likely transmission links and capacities in neighboring countries.

The impact of the free-trade environment on the nuclear phaseout is extremely small. In the high-electricity-price case, the costs of a nuclear phaseout are virtually identical to the base case, while in the low-price case a nuclear phaseout is a tiny bit (2%) less costly. This result is not surprising: With high-cost imports, Sweden would generate any replacement electricity domestically and ignore the import option. In the low-price case, Sweden would import its electricity rather than generate electricity domestically at the higher domestic price. But in either case, the trade would occur with or without a nuclear phaseout. Hence, the impact of the deregulation is minimal.[4]

Longer Lifetime for Nuclear Plants

There is obviously a large number of different possible sensitivity analyses that can be performed, but we have limited them to ones that seem to be most important for the nuclear phaseout costs. An important technical possibility, discussed above, is that the useful lifetimes of the nuclear plants will be longer than the forty years currently envisioned. Some sources suggest that a fifty-year lifetime is not unreasonable for a well-maintained plant, so we have estimated the cost of a phaseout in 2010 assuming that the lifetime of the average nuclear plant is to 2030 rather than 2020 on average.

The results are to raise the cost of the nuclear phaseout from $8.9 to $14.4 billion, for a 62% increase in the cost of the phaseout. Relative to the base case, this cost differential reflects a doubling of the length of the phaseout, offset by the effect of discounting the second decade of operation.

Expanding Hydroelectric Generation

A further genuine possibility for expanding electricity generation in the wake of a nuclear phaseout would be to expand the production of hydroelectric power. As Table 7 shows, there is theoretically still considerable room for expansion of hydroelectric power. However, the Swedish Parliament has decided to exclude the four major untouched rivers of the north from exploitation by hydroelectric power. Assuming that those four rivers are protected, the potential economical production is only 69 tWh versus normal production

Table 7. Potential Expansion of Hydroelectric Power.

Case	*Potential production level (tWh per year)*
Technical potential	130
Economic potential	90
Economic potential excluding the protected rivers	69
Parliament's plan for hydroelectric power	66
Current production level	63

Source: NUTEK.

today of 63 tWh and a future capacity of 67.8 tWh assumed in the SEEP model.

Even if there were a reversal of the decision not to exploit the protected rivers, there would be little impact of hydroelectric expansion on the cost of a nuclear phaseout. Hydroelectric power is extremely capital intensive. Indeed, it is possible that the existing hydroelectric capacity is overextended because of the effective subsidy to Vattenfall that is conveyed by the low cost of capital. NUTEK calculates that expansion of hydroelectric capacity would cost about $1714 per kWe (kilowatts of electrical energy), which at a 6% cost of capital would lead to a cost of electricity of 3.6 and 5 cents per kWh (in 1995 prices).[5] This compares with an estimated cost from gas of 5 cents per kWh calculated above (in 1995 prices). At a higher cost of capital, hydroelectric power would be uneconomical.

To calculate the impact of a hydroelectric capacity expansion, we assume that 20 tWh per year were available at a cost of 4.3 cents per kWh. This is an extremely optimistic assumption, for 20 tWh is the approximate upper limit of additional economic potential, *including* the four rivers that are currently off-limits. This would be slightly advantageous from a pure economic point of view, but it would lower the cost of a nuclear phaseout by only 2%. The reason for this is that hydroelectric power is barely a competitive technology if its costs are calculated with a realistic cost of capital.

One major reassuring conclusion, then, is that there is no hydroelectric dilemma. Even neglecting the intangible benefits of protection, the cost to Sweden of protecting its rivers from further hydroelectric development is relatively small, and in a world with a high Swedish cost of capital it is probably negligible.

Alternative Climate-Change Policies

My earlier analysis (see Table 1 in Chapter 5) showed that there is a significant interaction between Sweden's nuclear policy and its climate-change policy. If Sweden phases out its nuclear power, then the goal of holding CO_2 emissions at 1990 levels will become virtually impossible. Given the strong interaction between nuclear policy and climate-change policy, this section examines some alternative courses of action that Sweden might pursue.

One course, which would worsen the dilemma, would be to adopt even more ambitious goals for CO_2 emissions reductions. Many governments have endorsed a reduction of 20% in CO_2 emissions from 1990 levels, and the Berlin/Conference of the Parties in 1995 agreed to consider reducing emissions rather than stabilizing them at 1990 levels. Sweden has already endorsed reducing its CO_2 emissions in the years ahead. One question, then, is the extent to which more ambitious climate-change policies would make a nuclear phaseout more costly. For this case, we have assumed that Sweden stabilizes emissions at 1990 levels between 1990 and 2000 and then reduces CO_2 emissions by 20% after 2000.

A second potential approach is to modify Sweden's climate-change policy to allow Sweden to purchase CO_2 reductions in other countries. Going under the technical name of "joint implementation," this approach recognizes that it is *global* emissions that determine the pace of climate change; a unit reduction in CO_2 emissions is equally valuable whether it is in Sweden or in Poland or Russia. Therefore, just as nations can increase their living standards by engaging in fruitful trade in grain or electricity, so can they increase efficiency by allocating the global emissions reductions in a cost-effective manner. By joint implementation, those nations that can reduce emissions most inexpensively do so first. Global efficiency requires equalizing the marginal cost of emissions reductions around the world.

In this second approach, which is labeled an *offsets policy*, Sweden would purchase emissions reductions in the global marketplace at a given price. It would as a consequence reduce its emissions only up to the point where emissions reductions were less costly than the global price. Given that there is not currently a market in CO_2 emissions, estimating the cost of emissions reductions is largely guesswork at this stage. However, model estimates indicate that over the next two or three decades, a policy of efficiently stabilizing emissions would lead to a price for CO_2 emissions in the range of $10 to $40 per ton of CO_2 (or $40 to $150 per ton of carbon). For the offsets run, we have therefore assumed that Sweden could buy or sell carbon emissions rights at $25 per ton of CO_2 (in 1995 prices).[6]

The effect of these policies is instructive. The policy of reducing CO_2 emissions by 20% after 2000 imposes considerable further costs of climate change, estimated to be about $22 billion more than the emissions-stabilization option. The *incremental* cost of a nuclear phaseout is approximately the same for both the two CO_2 emissions-reduction policies, with a cost of slightly above $17 billion for the nuclear phaseout in each case, or about double the cost of the nuclear phaseout in 2010 without a climate-change policy. The offsets policy, by contrast, is much more favorable than either of the other two climate-change policies. The total cost of the nuclear phaseout plus CO_2 limitation is $53 billion without an offsets policy and $21 billion with an offsets policy. The additional cost of the nuclear phaseout is $11 billion with an offsets policy, this being about one-quarter higher than the cost of a nuclear phaseout with no climate-change policy.

We can also examine the climate-change policies in terms of the marginal cost of the last unit of CO_2 emissions reduced. These are shown in Tables 2 and 3 (pages 123 and 124) as the externality prices of CO_2 emissions. In the offsets case, we have set the cost of purchasing offsets at $25 per ton of CO_2. This compares with estimates of the damages from CO_2 emissions of between $2 and $10 per ton of CO_2 emissions in current assessments.[7] In the emissions-limitation cases, the marginal cost of CO_2 emissions reductions is between $112 and $135 per ton of CO_2 emissions reduction. Clearly, for Sweden the cost of limiting CO_2 emissions is extremely high and is much higher than current estimates of the estimated global damage from climate change.

These experiments suggest that putting the additional constraint of climate change on top of the nuclear phaseout will increase the cost of both constraints. However, there is a clear advantage to pursuing a policy in which reductions in CO_2 emissions can be efficiently allocated among countries rather than having each individual country pursue the same policy independent of its history, economic situation, and energy-market structure.

Summary and Discussion of Alternative Runs

Tables 5 and 6 show the results of the alternative runs of the SEEP models for different policies and different parametric assumptions. The impression from these runs that is the cost of a nuclear phaseout in 2010 without climate-change policies is robustly estimated to be between $7 billion and $14 billion (in 1995 prices and discounted to 1995). Moving the average date forward from 2010 adds approximately $1.4 billion per year for every additional year of early retirement. The different cases are also shown in Figure 2.

One important question concerns the trade-offs between nuclear policy and other national objections. In terms of interaction with other environmental goals, there is no significant conflict between the nuclear phaseout and the goal of protecting the four

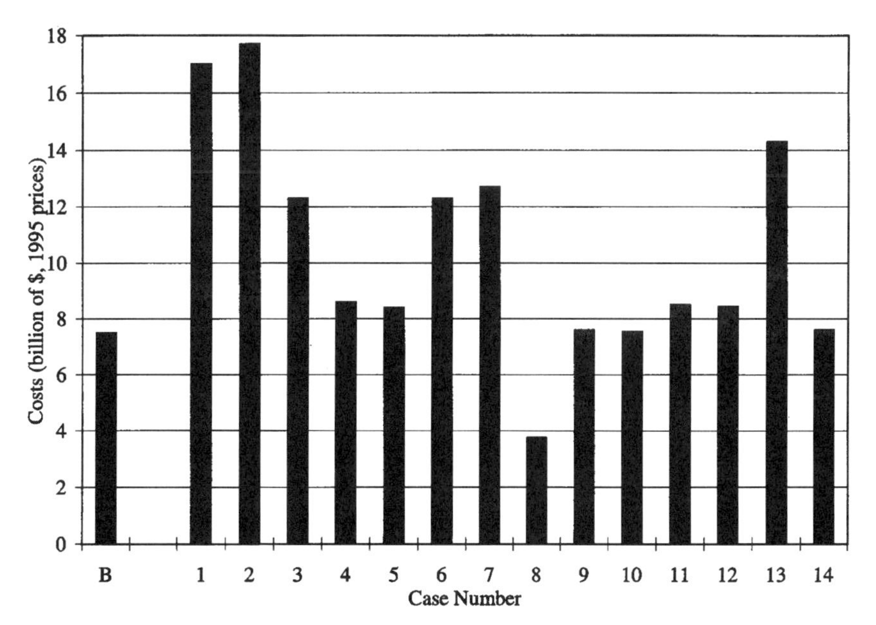

Figure 2. Impacts for Sensitivity Classes.

Note: See Table 6 on page 130 for the identification of the sensitivity run case numbers.

remaining rivers. Nor does the policy of electricity deregulation have a major impact on the wisdom of a nuclear phaseout. There is, however, a major and unfavorable interaction between the nuclear policy and climate-change policy, because limiting domestic CO_2 emissions increases the cost of a nuclear phaseout markedly.

It is important to understand that these estimates refer to the cost of a program discounted to 1995. The costs as of the date of phaseout will be considerably larger. This point was illustrated in Table 2 of Chapter 4 and must be reiterated. A 2010 phaseout costs $8.9 billion when discounted to 1995; but it will represent a larger economic burden when the phaseout actually occurs. In the base case, $8.9 billion in 1995 represents $18.4 billion at the date of the phaseout. Hence, the effect of discounting is like the effect of perspective in viewing distant objects in reducing the apparent size of the effect; for the case of a 2010 phaseout and a 5% discount rate, the cost will be approximately twice as high when the day actually arrives.

Note as well that the SEEP model provides estimates that are virtually identical to the simple supply-and-demand model described above for which estimates were shown in Table 2 of Chapter 4. There are two differences, working in opposite directions, that will produce divergent results in the two estimates. First, in the SEEP model, the market prices are somewhat below the cost of replacement power in the early part of the phaseout. For this reason, the estimated cost in the SEEP model will be somewhat below that in the simple supply-and-demand model. Second, the SEEP model includes costs of decommissioning, which add to the cost of the phaseout. In reality, these two factors just about balance each other out, so the estimated cost of the base program in the SEEP model is virtually identical to the theoretical supply-and-demand model.

ELECTRICITY PRICES

Most market participants will perceive the impacts of nuclear power policies in terms of the prices that they face for electricity and other goods and services. What is the likely evolution of prices under different policies? We have already noted that the impact of a nuclear phaseout will be to increase the price of electricity relative to a nuclear-continuation policy. Some people are surprised to learn that this increase in electricity prices will be felt mainly in the period after the phaseout but that there is unlikely to be a major impact on Swedish electricity prices in the long run. In the longer term, particularly in a deregulated market, prices will be determined by the marginal cost of new sources of electricity, which is largely independent of the decision about current nuclear power. This tendency will be reinforced if electricity market deregulation in Europe leads to extensive international sales and electricity price equalization in European countries. While the statement about the independence of long-run prices to the nuclear decision must be qualified, it is one of the major conclusions of both a theoretical analysis such as the one presented earlier in Figures 1 and 2 in Chapter 4, as well as the quantitative assessments like the SEEP model.

It should not, of course, be concluded that because the electricity prices are unlikely to rise in the long run that nuclear policy is irrelevant or costless. An analogy can make this point: After we have wrecked our car in an automobile accident, the cost of driving per kilometer may be the same as before the accident. But this does not imply that the accident is costless. Replacing a fine and workable car can be an expensive affair; similarly, even if electricity prices in Sweden eventually rise to a given level, the road to that price level can be more or less expensive in terms of resource costs.

What is likely to be the price trajectory with a nuclear phaseout? Figure 3 shows the calculated trajectory of industrial electricity prices with a nuclear phaseout that takes place linearly over the 2000–2010 period; this path is compared with the baseline price. According to the SEEP model, we expect the electricity price to rise gradually over the next two decades as electricity demand grows and capacity is more and more intensively used. According to the baseline projections, industrial prices would reach the level of replacement power plus distribution costs of about 7.2 cents per kWh shortly after the year 2010.

The expected impact of a nuclear phaseout would be for electricity prices to rise more quickly than in the baseline case. As nuclear power is being gradually phased out over this period, the price under the "phaseout" scenario rises, and it then reaches the cost of replacement power in 2003 rather than in 2014. At the maximum, the difference between the phaseout and baseline industrial

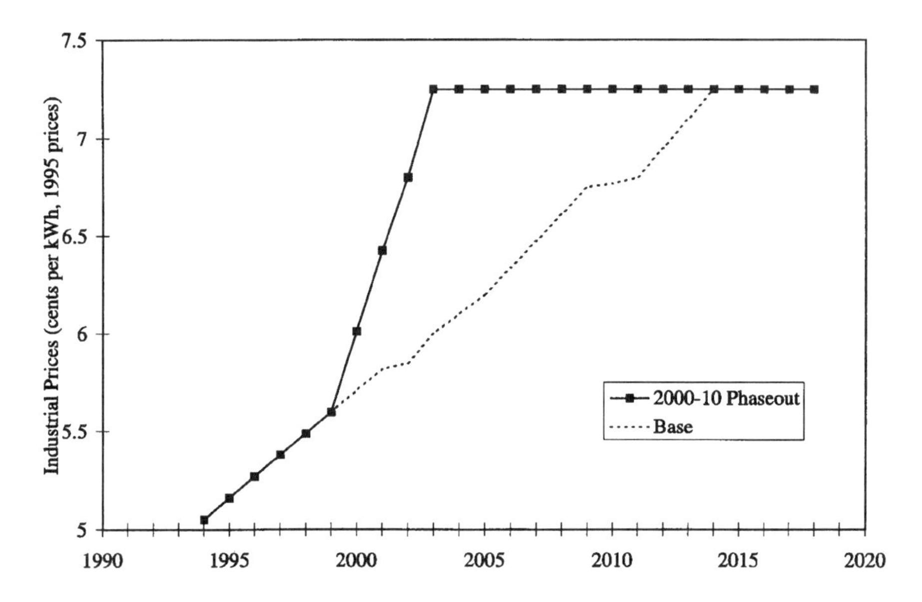

Figure 3. Prices with and without Nuclear Phaseout.

prices is about 25% in 2003. In this simplified analysis, we can see that the major impact will be a sharp rise in power prices in the next decade, followed by a stabilization of prices after 2010 at the new and higher cost of replacement power.

Reality will, of course, be more complicated than the model projections. Superimposed upon the smooth trends shown in Figure 3 will be influences of such things as rainfall, the business cycle, the ups and down of fuel prices, and the vicissitudes of operations in different plants. In addition, the actual numbers will differ from these projections insofar as the actual costs of production of new plants differ from our estimates. Moreover, if there are construction delays, if planning is imperfect, or if expectations are off base, there might be temporary spikes in the electricity price when nuclear plants are phased out.

A further consideration is that the costs of new plants may evolve over time and show some hysteresis as companies enjoy learning by doing or as the most favorable sites for construction are exhausted. Nor can we neglect the impact of the broader trends in industrial organization of the European electricity industry, whose price trends will spill over into Sweden in direct proportion as the European electricity industry is integrated and deregulated. And it is small consolation for those facing steep increases in electricity prices that these are ones that would have occurred otherwise—to learn that an economic harm is inevitable makes it no less unpleasant.

But these qualifications are likely to be of minor significance compared with the major impact of a nuclear phaseout. The essence of the impact is clearly shown in Figure 3, which shows the difference between the two price curves that opens up at the beginning of the phaseout and then gradually disappears as new capacity comes on line. This is the major message about electricity prices that this study has uncovered.

IMPACTS BY INDUSTRY

For the most part, we have analyzed the impact of a nuclear phaseout on the Swedish economy as a whole rather than on individual sectors. In this section, we explore briefly the potential impacts on individual sectors, focusing on the rise in electricity prices and the impact of that rise on production costs.

We have shown that, depending on the scenario, there is likely to be a substantial increase in electricity prices in the period following a nuclear phaseout. Industrial electricity prices are, as we showed above, low relative to many other major Organisation for Economic Co-operation and Development countries. As replacement power comes on line and as the price rises toward the marginal cost of replacement power, it is likely that the industrial electricity price will increase from a level around 4 cents per kWh in 1990 to a level around 7 cents per kWh (all in 1995 prices). The actual path will depend on the particular policies and conjunction of energy-market developments that unfold. Prices are likely to rise in any case, but they will rise more sharply if a nuclear phaseout takes place more quickly.

To estimate the impact of rising electricity prices on individual industries, I performed a simple calculation—asking about the impact of a *doubling of electricity prices* on production costs. To perform this calculation, I use standard cost theory and combine that with the 1985 Swedish input-output table. I then calculate the impact of rising electricity prices on production costs taking into account both the direct and indirect inputs of electricity to production in each sector.[8]

A brief explanation of the methodology will describe the approach. Economic life is a circular flow in which outputs of one process are inputs in another. Say that we are interested in determining the impact of rising electricity prices on export-intensive industries like specialty steel. The costs of production of steel will rise as electricity prices rise, of course. But steel includes other inputs, like transportation or iron ore, that also require electricity. And the flow is made even more complicated because electricity itself requires steel. In calculating the impact of electricity prices on steel, therefore, we need to include the indirect costs on iron ore and other inputs as well as the direct impacts on steel costs.

To perform this calculation requires manipulating and inverting an input-output matrix, which is easily accomplished thanks to modern computers. For this calculation, I have assumed that *only the electricity price changes* (there is, therefore, assumed to be no wage response or impact on other fuel prices) and that prices move proportionately with the cost of production. Tables 8 and 9 show the industries least and most affected by the hypothetical electricity-price doubling. The most heavily affected industries are ferroalloys manufacturing (15.4% increase), fiberboard manufacturing (9.8% increase), and nonferrous metals (7% increase). At the other end of the scale are fishing (0.3% increase), house construction (0.3% increase), and petroleum refining (0.4% increase).[9]

In addition, we can calculate the impact on the prices of major components of output by weighting the price increases in each category by the share of that industry in each component. The results of this calculation are shown in Table 10. The impact of the electricity-price doubling on the fixed-weight GDP deflator is just a hair under 2%. Because exports are electricity intensive, the price of exports will rise somewhat more than the general price level. The prices of import-competing goods and services are expected to rise somewhat less that the overall price level because these industries are slightly less electricity intensive than output as a whole. Consumer prices are expected to rise slightly less than the prices of the other components of output. Since it is assumed that wages are unchanged, the increase in the consumption price index is the decline in real wage rates.

Table 8. Total Price Changes in the Ten Sectors Least Affected by a 100% Increase in the Electricity Price (percent).

Sector	*Price increase*	*Share of total production*
Fishing	0.31	0.06
One- and two-family houses & leisure houses	0.31	6.49
Petroleum refining	0.40	1.25
Forestry and logging	0.63	1.26
Clothing	0.76	0.14
Tobacco manufacturing	0.82	0.12
Instruments & photo equipment	0.89	0.66
Electronics & telecommunications	0.90	1.48
Textiles, other than clothing	1.03	0.13
Wooden building materials	1.03	0.76

Table 9. Total Price Changes in the Ten Sectors Most Affected by a 100% Increase in the Electricity Price (percent).

Sector	*Price increase*	*Share of total production*
Ferro-alloys manufacturing	15.43	0.03
Fibreboard manufacturing	9.77	0.03
Nonferrous metals	7.00	0.29
Paper & board manufacturing	6.90	2.14
Iron & steel casting	6.46	0.08
General chemicals	6.37	0.65
Iron ore mining	6.10	0.18
Pulp manufacturing	5.71	0.69
Nonferrous ore mining	5.65	0.12
Water supply, sewage disposal	5.34	0.50

Table 10. Impact of a Doubling of Electricity Prices on Major Price Indexes.

	Total GDP	*Exports*	*Import-competing*	*Consumption*
Increase in price index	1.98	2.46	1.89	1.65

Note: This table show the impact of a doubling of electricity prices on price indexes of major components of output. This assumes that the price increase equals the increase in costs and that the electricity prices rise by 100%. In each case, the price increase for a given industry is multiplied by the share of that industry in the component.

How should we interpret these increases in production costs that arise from the likely coming increase in electricity prices? To begin with, as we have emphasized, the increases are probably inevitable in any case over the next two decades. The major uncertainty and impact of policy are the timing and degree of predictability of the electricity price increases. Complete deregulation and integration of the European electricity market (an unlikely possibility in the near term) would raise Sweden's domestic prices and, if other countries' policies were unchanged, Sweden would export electricity rather than products in which electricity is embodied. A nuclear phaseout would raise prices and would lead to electricity imports rather than exports, but the electricity-intensive exports would suffer without any compensating increase in electricity exports.

Those industries that experience sharp increases in relative costs because of the rising electricity price will find themselves at a competitive disadvantage at home and in international markets. This competitive disadvantage would arise whether the rising price comes from deregulation or nuclear phaseout. There are, fortunately, only a few major industries where the electricity price increase will have a substantial effect on relative costs. The heavily affected sectors in Table 9 comprised about 15% of Swedish exports in 1985. Of these, paper and board manufacturing are the dominant exports, but mining will also be heavily affected.

What will be the reaction in these industries? Unless these industries can engage in energy conservation or find low-cost fuels to substitute for electricity, they face a major profits squeeze in the coming years. They may move to a less energy-intensive product line. In addition, they may substitute other inputs that have risen less in price for electricity, and the effects will be less severe than those estimated here.[10] But where the product line is fixed and no other substitutes are readily available, the cost increases may be a major factor in an industry's competitiveness. It is good to recall through all this, however, that although we project a large price increase over the next two decades, the ultimate level of the electricity price will be largely independent of the nuclear phaseout. Hence, the industries that are severely affected will not be spared the cost increase if a nuclear phaseout does not occur. They will instead have a longer breathing spell for adjustment.

ENDNOTES

[1]The meaning of this assumption is that plants are built with realistic expectations about prices and energy and environmental policies and regulations.

[2]Surveys by the Nuclear Energy Agency indicate that most of the escalation in electricity generation costs occurred before 1985. In the latest survey of electricity-generation costs for 1992, there was little change in the capital costs from the prior survey. See Nuclear Energy Agency, *Projected Costs of Generating Electricity: Update 1992,* OECD, Paris, 1993.

[3]The annualized loss takes the present value of the loss as of the date of the phaseout and converts it into a perpetual annuity at the real cost of capital. Hence, the baseline loss of $8.9 billion converts into a lump sum of $18.4 billion at the time of phaseout, and that, in turn, is the equivalent of a $0.93 billion per year real annuity at a real discount rate of 5% per year. The higher discount rate leads to a higher annuity rate.

[4]Using a model of the Swedish electricity system developed by Lars Bergman and Bo Andersson ("Market structure and the price of electricity: an ex ante analysis of the deregulated Swedish electricity market," Stockholm School of Economics, mimeo, September 1994), the authors estimated the impact of a nuclear phaseout and concluded that the economic impact of a nuclear phaseout "is about the same in both a regulated and a deregulated environment" (Personal communication from Andersson, September 1, 1995).

[5]See NUTEK, *Scenarier*, p. 80.

[6]These results are drawn from the preliminary findings of EMF-14 discussed above. It should be noted that this price is above the estimated "optimal" price that would balance incremental costs with incremental benefits. That price is estimated to be about $10 per ton of carbon or $3 per ton of CO_2. For estimates of the optimal price, see William D. Nordhaus,

Managing the Global Commons: The Economics of Greenhouse Warming, MIT Press, Cambridge, Mass., 1994. A recent survey of the subject by the IPCC Working Group III finds somewhat higher estimates. See Intergovernmental Panel on Climate Change, Working Group III, James P. Bruce, Hoesung Lee, and Erik F. Haites, eds., *Climate Change 1995: Economic and Social Dimensions*, Cambridge University Press, New York, 1996, Chapter 6.

[7]For a discussion, see William D. Nordhaus, *Managing the Global Commons: The Economics of Climate Change*, MIT Press, Cambridge, Mass., 1994.

[8]The technique is a standard one and can be found in references on input-output analysis.

[9]These estimates may actually appear low relative to other figures. For example, one analyst pointed out that a doubling of electricity prices would lead to an increase by 22% of direct costs in the steel industry. This number differs from the present calculations, which indicate an increase of 6%, because the lower number includes all of value added in the denominator; the present calculation therefore includes not just direct costs but other elements of value added such as overhead labor and return to capital. In the long run, these indirect costs should be considered in weighing the impact of cost increases on prices.

[10]The input-output technique for estimating costs will generally give an upper limit to the increase in costs because it assumes fixed input-output coefficients. If the inputs of electricity can be substituted by other inputs (such as fuels that do not experience a price change), then the actual cost estimates will be lower than those shown here. For example, if production follows the log-linear or Cobb-Douglas production function, then the estimated cost increases shown in the tables are only 0.69 [(= ln(2)] times the values shown because of substitution. For smaller electricity price increases, the cost increases will be proportional to the ratio of the price increase to 100% times the values shown in the tables. The substitution effect will be roughly quadratic in the price increase.

PART III

The Future of Nuclear Power in Sweden

8

Resolving the Swedish Dilemma

Life is full of quandaries inside choices wrapped in dilemmas. The Swedish nuclear dilemma is a fine example of the tug-of-war between past commitments and present realities or between lofty goals and harsh realities. Up to this point, I have presented a technical analysis of the environmental and economic consequences of alternative courses of action. This final section draws the different strands together and considers further issues that cannot easily be quantified and modeled in a neat and tidy fashion.

The major issue raised by the Swedish nuclear dilemma is *the overall economic impact* of different approaches to Swedish nuclear power. The best estimate is that the cost of a nuclear phaseout in 2010 will be around $8.9 billion when discounted to 1995 (all figures given in this section are in 1995 prices). If costs are reckoned at the date of the phaseout (2010 in the base case), the costs are $18.4 billion.

Moving this forward by five years, so that the phaseout takes place evenly over the 2000–2010 period, will increase the costs to about $15 billion discounted to 1995. It seems unlikely that the phaseout can be moved much sooner than the straight-line 2000–2010 phaseout without risking temporary shortages with price spikes over the next decade.

How big is this cost in the context of the Swedish economy? On a per capita basis, and taking $18.4 billion at the time of the phaseout as a benchmark estimate, it is equivalent to reducing the stock of wealth by $2100 per person. Swedish gross domestic product (GDP) in 2010 is estimated to be around $340 billion, so the total cost is around 5% of the 2010 GDP. In overall economic terms, a nuclear phaseout will hardly bankrupt the nation, but it represents a loss in real income and wealth that is substantial by any account. This calculation represents the rock-bottom, irreducible minimum cost that would be imposed on Sweden by a nuclear phaseout in 2010.

The estimated costs will almost surely be larger if we take into account the interaction of the nuclear policy with other policies and if we recognize the realistic "frictions" that govern a modern industrial economy. Recall that the general equilibrium mode—the Swedish Energy and Environmental Policy (SEEP) model—used to estimate the costs of a nuclear phaseout assumes smoothly functioning markets with perfectly and rapidly adjusting wages and prices. To the extent that actual markets deviate from this idealized state—and there is little doubt that they do—the overall economic costs may increase relative to the costs estimated above. We might call the deviations of reality from our idealized world the macroeconomic, fiscal, and international frictions that tend to amplify underlying costs.

With respect to *macroeconomic frictions,* the most significant is probably the downward rigidity of wages prevalent in Sweden and other European countries. It is well established that because of expectations, inertia, and trade unions, nominal wages tend to rise to compensate for increased prices or lower real incomes. To the extent that a nuclear phaseout lowers real incomes or increases prices, therefore, real wage rigidity may magnify the costs. The extent of magnification will depend on the rigidity of real wages and on the extent to which monetary policy refuses to accommodate price increases.

There is no consensus among macroeconomists about the extent to which such frictions and rigidities can magnify the costs of microeconomic shocks. Nonetheless, the oil shocks of the 1970s were vivid testimony to the macroeconomic costs of energy shocks. Moreover, if Sweden were part of a monetary union at the time of the phaseout and price increases, the costs would be significantly higher than if inflation were allowed to increase in a transitory fashion because there would be no room for a rise in the general level of prices in Sweden. Nonenergy costs would have to decline relative to the baseline, and this would require a decline in output and employment to effect the deflation. Given the potential for sharp medium-term price consequences of a nuclear phaseout, it would be wise to include an additional cost for macroeconomic losses in the overall economic calculus.

A rigorous estimate of the macroeconomic frictions is beyond the scope of this study, but a simple calculation indicates that real wage rigidity can easily amplify the costs of a nuclear phaseout. Assume that a nuclear phaseout incurs a cost of 5% of one year's GDP, and that this cost is centered on 2010. Further assume that the "real-wage sacrifice ratio" is five. This indicates that to lower real wage by 1% requires lowering output by 1% of GDP for one year.[1] Finally, assume that monetary policy is completely nonaccommodating (say, because Sweden were part of a monetary union), so that maintaining external equilibrium would require sufficient unemployment to prevent any rise in wages. Under these assumptions, a nuclear phaseout would lead to an additional loss in output of about $41 billion (in 1995 U.S. dollars discounted to 1995). Under this calculation, the macroeconomic frictions would be four times as large as the rock-bottom resource costs that we estimated in the SEEP model.

It should be emphasized that alternative specifications will give quite different numbers. One alternative would be to assume that the major macroeconomic friction is nominal wage rigidity rather than the real wage rigidity just assumed. This approach leads to much smaller levels of macroeconomic costs because there is estimated to be no durable impact of the nuclear phaseout on the price level. While these calculations are full of uncertainties, they do suggest that the macroeconomic frictions due to inflexible wages and prices might add substantially to the costs of a nuclear phaseout.

A nuclear phaseout involves *fiscal consequences and frictions* as well. To begin with, consider the fiscal consequences for the Swedish government. The legal and fiscal consequences of a nuclear phaseout are unclear and remain to be determined by the Swedish Parliament and the courts. To a first approximation, however, it would seem reasonable to assume that the economic costs of the phaseout would become a government liability. There would be a direct revenue loss in the case of government-owned nuclear power stations (that is, for Vattenfall); here, the government would suffer a revenue loss equal to its prorated share of the estimated economic loss.

For privately owned nuclear power stations, the fiscal impact of the phaseout is unclear. If the government were to compensate the owners for the loss of profits, the estimated impact on the government debt of both publicly and privately owned plants would be the estimated loss as of the time of the compensation. For a 2010 phaseout, this would represent 5% of that period's GDP; for a 2005 straight-line phaseout, the required compensation would be about 7.5% of the 2005 GDP.

Sweden is presently in the throes of a major fiscal crisis, with a central government deficit averaging around 10% in 1994–1995. The government is currently taking extraordinary steps to decrease the fiscal deficit, and it is obvious that a measure that would increase the debt-GDP ratio by 5 to 7.5 percentage points would be most unwelcome. The effects of a higher government debt are not, of course, purely accounting ones. Sweden will either have to raise tax rates above their already high levels or further squeeze expenditures to compensate for the higher debt. Here again, the frictions or efficiency losses involved in tax increases or expenditure decreases will tend to amplify the estimated costs. For example, some studies suggest that the marginal cost of raising one additional krona of taxes might be an additional 20% to 50% of revenues in efficiency losses. Here again, as in the case of macroeconomic frictions, our calculations are likely to understate the true cost of a nuclear phaseout.

One major consequence of a nuclear phaseout will be a greater dependence on foreign sources of energy, and this raises the issue of *international repercussions and frictions*. Like the other frictions, the international impact is extremely complicated and rife with economic controversies, but the range of possibilities can be easily sketched. Consider the direct impact on energy purchases. To replace the 70 terawatt-hours of nuclear electricity, a minimal impact would come with replacement by domestic generation using coal. At a coal price of $50 per ton, this would lead to an annual import bill of around $1.4 billion; other fossil fuels would lead to greater import costs. At the other extreme, say that the entire replacement electricity were imported at a price of 5 cents per kilowatt-hour; this would lead to an annual import bill of about $3.5 billion. Therefore, a reasonable range of estimates is between $1.4 billion and $3.5 billion per year of increased energy imports from a nuclear phaseout. Of course, these figures are subject to the usual uncertainties about future price movements, which are particularly large for oil prices, and therefore to any gas purchases where the prices are linked to oil, but the uncertainties are probably also quite large for the price of imported electricity.

Compared with current imports of around $57 billion in 1994, these numbers are substantial. One should not, of course, fall into the trap of energy mercantilism, wherein the only good energy is produced domestically, but one must recognize that adding 2.5% to 6% to a nation's import bill is no small decision.

The real costs of the additional imports actually come when the real adjustment of international trade to the increased imports takes place, and this leads to what we might call the *international-trade friction*. The higher energy imports are likely to be paid for by a decline in Sweden's real exchange rate—either through a real depreciation in the krona or through a decline in the prices of domestic output relative to foreign production. This decline in the real exchange rate will be necessary to decrease nonenergy imports and increase exports to pay for the higher energy imports. The change in the terms of trade will impose additional costs on top of the real resource costs calculated above. In a sense, these are the international equivalent of the macroeconomic frictions discussed above and would apply to any country that is not a complete price taker in world markets.

The size of these additional real income losses is speculative, but illustrative calculations based on realistic import and export elasticities might lead to an additional cost on the same order of magnitude as the size of the increased imports. On a present value basis, then, the international-trade frictions might be about as large as the rock-bottom estimate of the cost of the nuclear phaseout.

To summarize the additional frictional costs of a nuclear phaseout, we estimate that the macroeconomic frictions from the price-raising impact comprise between 0% and 400% of the rock-bottom resource costs of a nuclear phaseout; the fiscal costs are likely to add between 20% and 50% to the costs; and the terms of trade effects are approximately the same magnitude as the resource costs. This is obviously a wide range of estimates, but it emphasizes that our estimated costs are likely to be on the low side of reality.

What about *health, safety, and environmental impacts*? The major purpose of the nuclear phaseout is to protect the health and safety of the Swedish population. We have reviewed the evidence concerning the relative safety of nuclear and other major sources of electricity above. The conclusion seems clear that for routine releases, nuclear power is likely to have smaller health, safety, and environmental impacts than the other major fuel cycles other than natural gas. The evidence is strong that coal-fired and oil-fired stations have significant adverse environmental and health effects. Of the fossil fuels, natural gas has the smallest health, safety, and environmental impacts, but introducing a major new fuel system into the Swedish economy will raise difficult and controversial issues of siting, pipelines, and even disposing of deep-sea facilities. The two remaining question marks about nuclear power are the risk of large-scale accidents and long-term waste storage. While these questions are unlikely to be completely resolved for many decades, current plans appear to have reduced the risks to levels that are low relative to routine risks in other fuel cycles.

Our conclusion with respect to environmental issues is that there is likely to be a deterioration in environmental quality in Sweden (or in neighboring countries) following a nuclear phaseout. There is, however, no hydroelectric dilemma: the goal of protecting Sweden's remaining unharnessed rivers does not interact in a significant way with nuclear policy.

This study has touched only briefly on the complex issues involved in *deregulation* of the Swedish and European electricity industries. Academic studies and experience in other countries indicate that a well-designed deregulation proposal can have substantial benefits in reducing the cost of electricity and "getting the prices right." Sweden seems headed down the right path in this respect with one exception, and that is the decision to leave Vattenfall as the dominant producer in the electricity industry. The decisions on deregulation do not, however, appear to have a significant effect on the economic impact of a nuclear phaseout—there is no "regulatory dilemma" in the nuclear decision.

Although Sweden is a small country producing but a tiny fraction of the world's energy and output, it plays a large role on the stage of world opinion making. In this respect, we might ask how its decisions affect *nuclear power in the global context.* The world is watching Sweden's redesign of the welfare state, and it will also pay close attention to how Sweden manages its nuclear power industry. Sweden is unique among democracies in having a well-managed nuclear power industry, producing at low cost, and having found a political resolution to the thorny issues of siting and disposal of nuclear wastes. How is the world to read the message of shutting this industry down in the prime of its economic life? Will this be a nail in the coffin of the nuclear industry? Will this doom attempts to develop advanced inherently safe nuclear power reactors? These are serious questions. Sweden should recognize that its actions have a demonstrable effect on other countries and that its actions will be read internationally as an important message about the viability of nuclear power as a future energy source.

The issue of future energy sources is vitally related to the question of the threat of future *climate change*. Just as Sweden's decision about nuclear power will be widely viewed as a judgment on the viability of nuclear power, so will it convey a message about the relative priority of domestic energy concerns and international environmental commitments. As the analysis above shows, Sweden will be hard pressed to keep its international commitments on stabilizing carbon dioxide (CO_2) emissions even with its nuclear industry producing full tilt. If Sweden phases out its nuclear power, then (unless some technological miracle occurs or some fancy emissions bookkeeping takes place) it will be virtually impossible for Sweden to keep its commitments on CO_2 emissions stabilization.

Here, Sweden faces a dilemma of commitment. How should it resolve the competing promises to the international community about climate change and to itself about the nuclear phaseout? The temptations for countries to renege on international environmental agreements is enormous given that the costs of compliance are local, concrete, and immediate, while the benefits are dispersed, diffuse, and in the distant future. If countries widely perceived as honorable and responsible in the international arena turn their backs on international environmental agreements, then small, poor, or unscrupulous countries will quickly follow their examples.

The other side of the commitment about climate change is that keeping both the nuclear and the climate-change commitments raises the price of both. We estimate that a nuclear phaseout in the year 2010 (discounted to the phaseout date of 2010) will cost $19 billion. If this policy is combined with a limitation of CO_2 emissions to 1990 levels, the price tag of both policies rises to $109 billion, or one-third of one year's GDP. If we add the macroeconomic, fiscal, and international-trade frictions to this number, the total price tag becomes frighteningly large.

At the same time, the astronomical cost of putting climate-change policy together with a nuclear phaseout may lead policymakers to see the wisdom in new approaches to climate-change policy. I point particularly to the use of cost-effective strategies in which emissions reductions are allocated among nations so that CO_2 reductions take place in those countries where the marginal cost of reducing emissions is lowest. If future climate-change policies lead to global emissions trading, so that Sweden can purchase carbon emissions from low-cost sources, then the added burden of its climate-change commitments is likely to be quite modest.

In the nuclear debate, we must, of course, consider the role of *political commitments* and deal with the question of how to treat the nuclear referendum of 1980. Should it be treated as a solemn, quasi-constitutional determination that binds future generations independently of future events, of the turnover of population and opinion, of new information, and of changing economic circumstances? Or is it more like a poll or election that reflects the opinions or fads of the electorate at a particular point in time but can be rejected if circumstances change? Is the current generation bound by a determination almost two decades old even while it uproots other commitments and legislative acts that were taken at that time or even later?

These are deep and difficult philosophical questions that political leaders in Sweden must grapple with, and no foreign observer can presume to give a definitive answer. But the following factor should be considered as Sweden weighs its options: Sweden is currently engaged in questioning and rethinking many of the fundamental assumptions of its political and economic system. This reexamination involves the degree of government involvement, the size of government, the tax and expenditure structure, and many other features of Sweden's social system that have been established over the last half century. It would seem ironic if all these fundamental parts of the Swedish social contract between its government and its citizens were reexamined while a single item—the nuclear power question—were not subjected to the same root-and-branch reconsideration.

The basic message of this study is that in light of the many years of further experience and investment since the 1980 nuclear referendum, the economic and environmental rationale for a nuclear phaseout is very thin. The direct economic losses from a phaseout are substantial, and these losses are likely to be amplified by macroeconomic, fiscal-policy, and international frictions. There are unlikely to be any gains in environmental quality, health, or safety from moving to a different fuel cycle for generating electricity, and some fuel cycles, particularly coal-based or oil-based ones, are likely to lead to a significant deterioration in health and environmental conditions. Sweden would be doing significant damage to the enhancement of rational decisionmaking if it turned its back on its international commitments and simply followed a path chosen in different circumstances fifteen years ago. In short, Sweden has much to lose and little to gain from an early retirement of its nuclear power industry.

The question of nuclear phaseout comes at a turning point in the postwar Swedish history. As Assar Lindbeck and his colleagues wrote in *Turning Sweden Around*:

> At some rare moments a nation pauses to reflect on its future. Such moments usually occur in periods of decline and crisis. The ability of a nation to reconsider past decisions and rejuvenate itself is then put to the test. Today Sweden is experiencing its most serious crisis since the 1930s. How Sweden works to overcome the crisis will mark the country for decades to come.[2]

Resolving the nuclear dilemma—reconsidering this past decision in a rational fashion—is just one part of the task of setting Sweden on a secure course over the next century.

ENDNOTES

[1]For a discussion of the sacrifice ratio and representative estimates, see N. Gregory Mankiw, *Macroeconomics*, Worth Publishers, New York, 3rd edition, 1997. Most calculations of the sacrifice ratio refer to nominal-wage rigidity, but it may be more appropriate for Sweden to deal with the real-wage rigidity that is prevalent in Europe.

[2]Assar Lindbeck et al., *Turning Sweden Around*, MIT Press, Cambridge, Mass., 1995, p. 1.

Index